CHAPTER ONE

1. **Evaluate** without using mathematical tables or calculators, the square root of

$$\frac{0.0273 \times 1.152}{1.3 \times 1.68}$$

2. **Find** the integral values of the following simultaneous linear inequalities.

$$\frac{x + 2}{2} \geq 5; \; \frac{x + 6}{4} < 4$$

3. A cold water tap can fill a bath in 10 minutes while a hot water tap can fill it in 8 minutes. The drainage pipe can empty it in 5 minutes. The cold water and hot water taps are left running for 4 minutes. After which all the three taps are left running. **Find** how long it takes to fill the bath.

5. The hire purchase term of a cupboard is a deposit of $44 and six monthly instalments of $9 each. The hire purchase price is 175% of the cost price while the cash price is 25% more than the cost price. **What** is the cash price of the cupboard?

6. The circle below whose area is 18.05 cm^2 circumscribes a triangle ABC where AB = 6.3 cm, BC = 5.7 and AC = 4.8cm. Find the area of the shaded part.

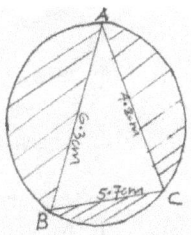

7. A teacher gave his form four class a quiz in mathematics which was marked out of 50 marks. The distribution of the marks was as shown in the table below.

Mark	10-14	15-19	20-24	25-29	30-34	35-39	40-44
Frequency	2	4	6	10	9	7	2

Calculate the median of this class.

8. The actual area of an estate is 3510 hectares. The estate is represented by a rectangle measuring 2.6cm by 1.5 cm on the map whose scale is 1:n. **Find** the value of n.

9. Wasike and Wanjala live 40km apart. Wasike starts cycling from his home at 8.00a.m toward's Wanjala's house at 16km/h. Wanjala stars cycling towards Wasike's house 30 minutes later at 8km/h. **what** time did they meet.

10. **Solve** for x in the equation

$$Log_8 \ (x+6) - Log_8(x-3) = \frac{2}{3}$$

11. The surface area of a spherical ball is increased by 21% after pressure was pumped in.
   (a) **Find** the circumference of the original ball if the one with increased pressure has a circumference of 55cm.
   (b) **Calculate** the percentage increase in volume of the ball.

12. The figure below shows speed time graph of a journey. If total distance travelled in 80 second is 920m

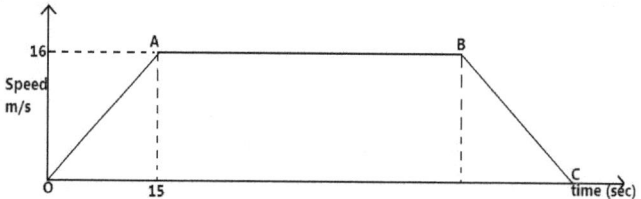

   i. **Calculate** the acceleration during the first 15 seconds.
   ii. The distance travelled in the final 40 seconds.

13. Use table of cubes and reciprocals to **evaluate**.

$$45.7^3 - \sqrt[3]{4411} + \frac{1}{0.07897}$$

14. Given the simultaneous equations:
   $$5x + y = 19$$
   $$-x + 3y = 9$$
   Write down the equations in matrix form hence find the values of x and y.

15. A man invests $240 in an account which pays 16% interest p.a. the interest is compounded quarterly. **Find** the amount in the account after 1 ½ years.

16. The figure shows a circle centre O with line POS as a diameter. QOR is an equilateral triangle and PT =ST. given that ∠ POQ =64⁰, **find** the sizes of ∠ SPT and ∠ STR.

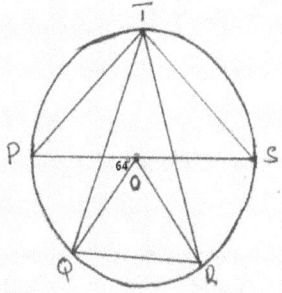

17. The figure below shows the graph of logy against logX

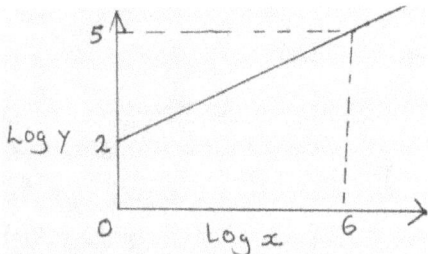

If the law connecting x and y is of the form y = ax^b, where a and b are constants. **Find** the values of a and b.

18. A straight line passing through the points (8,-2) and (4,-4) has its equation in the form ax + by + c = 0, where a, b and c are integers.
a) **Determine** the numerical values of a, b and c.

b) If the line in (a) above cuts the x-axis at point P, **determine** the coordinates of P.

c) Another line, which is perpendicular to the line in (a) above passes through point P and cuts the y axis at Q. **determine** the coordinates of point Q.

Find the length of QP

19. A helicopter is stationed at an airport H on a bearing of 060⁰ and 800km from another airport P. A third airport J is on a bearing of 140⁰ and 1200km from H.

a) Using a scale of 1cmrep. 100km.

(i)   **Show** the relative positions of P, H and J
(ii) **Determine** the distance between P and J.
(iii) State the bearing of P from J.

b) A jet flying at a speed of 1.035Km/hr left J towards P. The helicopter at H also took off towards P at the same time. Find the speed at which the helicopter will fly so as to arrive at P, 12 minutes later than the jet.

20. **Using** a ruler and a pair of compasses only construct.
      a) A line PQ, 8cm long. On the line construct triangle PQR such that $\angle$ QPR = 75^0 and line PR = 7cm. measure line QR
      b) **Construct** a circum circle of triangle PQR and measure its radius.
      c) **Calculate** the difference in area between the circle and triangle PQR.

21. A two digit number is such that its value is equal four times the sum of its digits. If the digits are interchanged, the new number formed exceeds two-thirds of the original number by 52. **Find** the original number.

22. The prices of admission to a concert are as follows:
Primary school children   - $1.00 each
Secondary students   - $2.00
University   - $4.00
One day the money taken from the university students was twice the proceeds of the primary sales while four times as many as tickets sold to secondary as to primary. If the total collections at the ticket office were $220 **find** the number of tickets which were sold altogether.

23. Four ships are at sea, such that ship B is 520km on a bearing of 210^0 from ship A. Ship C is due North of ship B and due west of ship A. The fourth ship D is 100km on a bearing of 340^0 from ship A and 240km on a bearing of 070^0 from C.
Draw a rough sketch showing the positions of ships A, B, C and D.
a) Use your sketch to **find** the size of
(i) angle ADC
(ii) angle BAD
b) **calculate** to the nearest whole number
(i) the distance of C from A.
(ii) the area of the quadrilateral ABCD.
(iii) determine the bearing of B from D

24. (a) Given that $y = 7 + 3x - x^2$, complete the table below.

X	-3	-2	-1	0	1	2	3	4	5	6	
Y	-11			7							-11

(b) On the grid provided and using a suitable scale, draw the graph of
$y = 7 + 3x - x^2$

© On the same grid draw the straight line and use your graph to solve the equation. $x^2 - 4x - 3 = 0$
(d) **Determine** the coordinates of the turning point of the curve.

4

25. A parallelogram has the lengths of its two diagonals being $4\sqrt{5}$ cm and $8\sqrt{5}$ cm. The acute angle made by the diagonals at the point of intersection is 60^0
(do not use mathematical tables and calculators in part (a) and (b) of this question)

Calculate
(i) the perimeter of the parallelogram
(ii) the angles of the parallelogram.
(iii) The area of the parallelogram

SOLUTIONS TO CHAPTER ONE

1.	$\dfrac{0.0273 \times 1.152 \times 1000}{1.3 \times 1.68 \times 1000}$ $\dfrac{27.3 \times 1.152}{13 \times 168}$ $\sqrt{0.0144} = 0.12$
2.	$x + 2 \geq 10 \Rightarrow x \geq 8$ $x + 6 < 16 \Rightarrow x < 10$ $\therefore$ int *egral values are* 8 & 9
3.	$\left(\dfrac{1}{10} + \dfrac{1}{8}\right) \times 4 = \dfrac{9}{40} \times 4 = \dfrac{9}{10}$ $\dfrac{1}{10} + \dfrac{1}{8} - \dfrac{1}{5} = \dfrac{4 + 5 - 8}{40} = \dfrac{1}{40}$ $\therefore$ *Time taken* $= \dfrac{1}{10} \div \dfrac{1}{40}$ $= \dfrac{40}{10} = 4\,\text{min}$
4.	$H.P = 41400 + \left(900 \times 6\right) = 9800 / =$ *Cost price* $\dfrac{100}{175} \times 9800 = 5600 / =$ $\therefore$ *cash price* $= \dfrac{125}{100} \times 5600 = 7000 / =$
5.	$S = \sqrt{S\,(S - a)(S - b)(S - c)}$ $S = \dfrac{1}{2}\left(4.8 + 6.3 + 5.7\right) = 8.4$ *Area* $= \sqrt{8.4\,(2.7)\,(3.6)\,(2.1)}$ *Area* $= \sqrt{171.4608} = 13.0943$ $\therefore 18.05 - 13.0943 = 4.9557$ ≈ 4.96

6.	$M = 24.5 + \dfrac{8.5}{10} \times 5$ $= 24.5 + 4.25$ $= 28.75$
7.	$3510 \times 10{,}000 = 35100000\ m^2$ $Model = 2.6 \times 1.5 = 3.9\,cm2$ $1cm^2 = \dfrac{35100000}{3.9} = 9{,}000000\ m^2$ $\therefore\ 1cm = \sqrt{9{,}000{,}000} = 3000\ m$ $hence\ \ scale : 1 : 3000 \times 1000$ $n = 3{,}000{,}000$
8.	$lex\ x\ be\ meeting\ \ dis\tan ce\ from\ Wasike$ $\Rightarrow \dfrac{x}{6} + \dfrac{1}{2} = \dfrac{40 - x}{8}$ $3x = 72 \Rightarrow x = 24\,km$ $time = 8.30\ a.m + \dfrac{24}{16}$ $= 8.30 + 1.30 = 10.00\ a.m$
9.	$\dfrac{x + 6}{x - 3} = 8^{\,2/3}$ $(x + 6) = \left(2^3\right)^{2/3} (x - 3)$ $x + 6 = 4\,(x - 3)$ $x + 6 = 4x - 12$ $-3x = -18 \Rightarrow x = 6$
10.	$(a)\ g\ .\ s.\ f\ \dfrac{121}{100} = 1.21$ $l.s.f = \sqrt{1.21} = 1.1$ $\therefore\ circumfren ce = \dfrac{55}{1.1} = 50\,cm$ $(b)\,Percentage\ \ increase =$ $\quad \left(1.1^3 - 1\right)100\%$ $\quad\quad = 33.1\%$
11.	$(a)\ \dfrac{\Delta\ speed}{\Delta\ t} = \dfrac{16}{5} = 1\dfrac{1}{15}\ m/s^2$

	(b) $16 \times 10 + \dfrac{1}{2} \times 30 \times 16$ $dist = 160 + 240 = 400 \; m$
12.	$45.7^3 = 9544$ $3\sqrt{4411} = 16.40$ $\text{Re}\,p\; 0.07897 = 12.66$ $\therefore = 9544 - 16.40 + 12.66$ $= 9540.26$
13.	$\begin{pmatrix} 5 & 1 \\ -1 & 3 \end{pmatrix} \begin{pmatrix} x \\ y \end{pmatrix} = \begin{pmatrix} 19 \\ 9 \end{pmatrix}$ $\dfrac{1}{16} \begin{pmatrix} 3 & -1 \\ 1 & 5 \end{pmatrix} \begin{pmatrix} 5 & 1 \\ -1 & 3 \end{pmatrix} \begin{pmatrix} x \\ y \end{pmatrix} = \dfrac{1}{16} \begin{pmatrix} 3 & -1 \\ 1 & 5 \end{pmatrix} \begin{pmatrix} 19 \\ 9 \end{pmatrix}$ $\begin{pmatrix} 1 & 0 \\ 0 & 1 \end{pmatrix} \begin{pmatrix} x \\ y \end{pmatrix} = \dfrac{1}{16} \begin{pmatrix} 48 \\ 64 \end{pmatrix}$ $\begin{pmatrix} x \\ y \end{pmatrix} = \begin{pmatrix} 3 \\ 4 \end{pmatrix}$ $x = 3 \; and \; y = 4$
14.	$p = 24000, r = 4\%, n = 6$ $\therefore AM = 24{,}000 \left(1 + \dfrac{4}{100}\right)^6$ $= 24000 \times 1.04^6$ $= 24000 \times 1.265319$ $= 30367.70$
15.	$\angle SPT = \dfrac{1}{2} \angle POQ$ $= \dfrac{1}{2} \times 64 = 32$ $\angle STR = \dfrac{1}{2} \angle SOR$ $= \dfrac{1}{2} \left(180 - (64 + 60)\right)$ $= \dfrac{1}{2} \left(180 - 124\right) = 28^0$
16.	$Log \; y = Log \; a + b \log x$ $\therefore \log a = 2 \Rightarrow a = 100$ $6b + 100 = 5 \Rightarrow b = 15 \dfrac{5}{6}$

17..	$\dfrac{y + 2}{x - 8} = \dfrac{-4 + 2}{4 - 8} = \dfrac{1}{2}$ (a) $2y + 4 = x - 8$ $\Rightarrow x - 2y - 12 = 0$ $\therefore u = 1, b = -2$ and $c = -12$ $y = \dfrac{x}{2} - 6$, when $y = 0$ (b) $\dfrac{x}{2} - 6 = 0 \Rightarrow x = 12$ $\therefore$ Coordinate s of pane $(12, 0)$ $\dfrac{y}{x - 12} = -2$ (c) $\Rightarrow y = 24 - 2x$ but when $x = 0$, $y = 24$ $\therefore$ Coordinate s of Q are $(0, 24)$ (d) $QP = \sqrt{(12 - 0)^2 + (0 - 24)^2}$ $= \sqrt{144 + 576} = 26.83$ Units
18.	(a) $PJ = 1560$ km (ii) $360^0 - 110^0 = 250^0$ $\dfrac{800}{x} - \dfrac{1560}{1035} = \dfrac{12}{60}$ (b) $\dfrac{800}{x} = 0.2 + 1.507$ $\therefore x = \dfrac{800}{1.707} = 468.66$ km / hr

19.	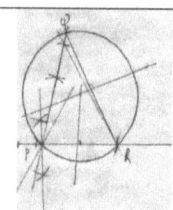
20.	(a) let the no. be xy $10(x) + y = 4(x+y)$ $3y = 6x$ $y = 2x$-------------(i) $10y + x - 2/3 (10x + y) = 52$ $28y - 17x = 156$-------------(ii) $\Rightarrow 28(2x) - 17x = 156$ $x = 4$ But $y = 2x \Rightarrow y = 2(4) = 8$ $\therefore$ original number $= 48$ (b) let primary proceeds be x university students be 2x $\Rightarrow$No. of tickets Primary $= \dfrac{x}{100}$. University $= \dfrac{2x}{400} = \dfrac{x}{200}$ Secondary $=$ $4\left(\dfrac{x}{100}\right) = \dfrac{x}{25}$ $\left(\dfrac{x}{100}\right)100 + \left(\dfrac{x}{200}\right)400 + \left(\dfrac{x}{25}\right)200 = 22000$ $x + 2x + 8x = 22000$ $x = 2000$ *Total no. of tickets* $\dfrac{x}{100} + \dfrac{x}{200} + \dfrac{x}{25} = \dfrac{2x + x + 8x}{200}$ $= \dfrac{2(200) + 200 + 8(200)}{200}$ $= 110 \ tickets$

21

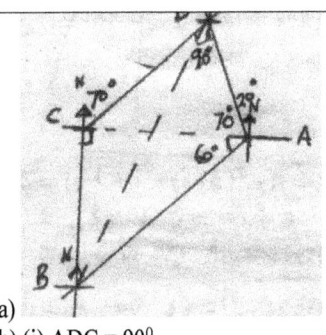

(a)

(b) (i) $ADC = 90^0$

(ii) $ACB = 90^0$

$$\frac{CA}{Sin\ 30^0} = \frac{520}{Sin\ 90^0}$$

(c) (i) $CA = \dfrac{52^0\ Sin\ 30^0}{Sin\ 90^0}$

$= 260\ km$

$$\frac{BC}{Sin\ 60^0} = \frac{520}{Sin\ 90^0}$$

(ii) $BC = \dfrac{520\ Sin\ 60^0}{Sin\ 90}$

$= 450\ km$

(iii) bearing of A from B is 030^0

22.

-2	-1	1	2	3	4	5
-3	3	9	9	7	3	-3

$x = -0.8$

$x = 4.6$

23	(a) $AB^2 = \left(4\sqrt{5}\right)^2 + \left(2\sqrt{5}\right)^2 - 2 \times 4\sqrt{5} \times 2\sqrt{5} \ Cos \ 60^0$.

(a) $AB^2 = \left(4\sqrt{5}\right)^2 + \left(2\sqrt{5}\right)^2 - 2 \times 4\sqrt{5} \times 2\sqrt{5} \ Cos \ 60^0$.

$= 80 + 20 - 40$

$AB^2 = 60 \Rightarrow AB = \sqrt{60} = 2\sqrt{15}$

$BC^2 = \left(2\sqrt{5}\right)^2 + \left(4\sqrt{5}\right)^2 - 2 \times 2\sqrt{5} \times 4\sqrt{5} \ Cos \ 120^0$

$= 80 + 20 + 40$

$BC^2 = 140 \Rightarrow 13c = \sqrt{140} = 2\sqrt{35}$

$\therefore Perimeter = 2\left\{\sqrt{35} + 2\sqrt{15}\right\}$

$= \left(4\sqrt{35} + 4\sqrt{15}\right) units$

$\dfrac{SinB}{4\sqrt{5}} = \dfrac{Sin \ 60^0}{2\sqrt{15}}$

$Sin \ B = \dfrac{Sin \ 60^0 \times 4\sqrt{5}}{2\sqrt{15}}$

(b) $SinB = 1 \Rightarrow B = 90^0$

$\Rightarrow \angle ABC = 90 + 40.89 = 130.89^0$

$\therefore \angle DAB = 360 - \dfrac{2 \times 130.89}{2} = 49.11^0$

(c) $A = \dfrac{1}{2} \times 2\sqrt{5} \times 4\sqrt{5} \times \dfrac{\sqrt{3}}{2} \times 4$

$= 40\sqrt{3} \ cn^2$

CHAPTER TWO

1. Use logarithms to **evaluate**

$$\frac{16.49^2 \times \sqrt[3]{0.6329}}{438.2}$$

2. **Solve** the equation below.

$$7^{2x} - 8 \times 7^x + 7 = 0$$

3. The gradient of a curve at any point is given by 2x-1. Given that the curve passes through point (1,5). **Find** the equation of the curve.

4. A car was valued at Ksh. 500,000 in January 2008. Each year, its value depreciates at 12% p.a. find after how long would the value depreciate to Ksh. 250,000

5. Given that $2 \leq A \leq 4$ and $0.1 \leq B \leq 0.2$. **Find** the minimum value of $\dfrac{AB}{A - B}$

6. A surveyor finds that she needs 28 beacons placed 40m apart when she surveys a length of the road. If she were to place the beacons 30m apart, **how many** would she need?

7. The first and thirteenth terms of A.P are 7 and 1 respectively. **Calculate** the number of terms which have a sum of zero.

8. The internal and external diameters of a circular ring are 8cm and 10cm respectively. **Find** the volume of the ring if its thickness is 3.5 millimeters.

9. Given that $Cos\, x = \dfrac{2}{\sqrt{5}}$. Without using tables or calculators **evaluate** and **simplify** leaving your answer in the form of $a\,\sqrt{b}$

(a)Sin x

(b)Tan(90 – x)
(1mk)

10. A two digit number is formed from the first four prime numbers.
(a) Draw the table to show the possible outcomes.

(b) Calculate the probability that a number chosen from the two digit numbers is an even number.

11. (a) **Expand** $(a-b)^5$

(b) Use the first three terms of the expansion in (a) in ascending powers of b to **find** the approximate value of $(1.98)^5$

12. The diagram below shows a circle ABCDE. The line FEG is a tangent to the circle at point E. Line DE is parallel to CG, $\angle DEC = 28^0$ $\angle AEG = 32^0$

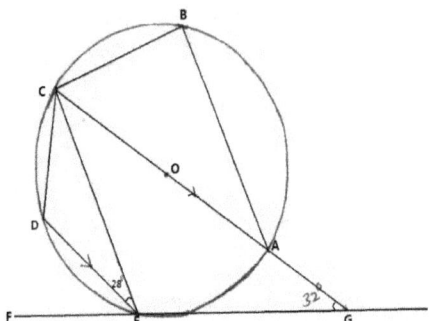

Calculate

(a) $\angle$ AEG

(b) $\angle$ ABC

13. The following distribution shows the masses to the nearest kilogram of 50 pupils in class 8.

Mass (kg)	26 – 30	31 – 35	36 – 40	41 – 45	46 - 50	51 – 55
Frequency	4	12	18	11	4	1

Calculate the standard deviation.

14. Make x the subject of the formula. $h = \sqrt[3]{\dfrac{c - x^2}{b}}$

15. The resistance of an electrical conductor is partly constant and partly varies as the temperature(k). when the temperature is 27⁰C, the resistance is 55 ohms, and when the temperature is 57⁰C, the resistance is 58 ohms. **Find** the relation between the temperature and the resistance.

16. A dam containing 4158m³ of water is to be drained. A pump is connected to a pipe of radius 3.5cm and the machine operates for 8 hours per day. Water flows through the pipe at the rate of 1.5m per seconds. **Find** the number of days it takes to drain the dam.

17. (a) A shear parallel to the x-axis (the invariant line) maps point (1,2) on to point (7,2). T is the transformation equivalent to this shear followed by the reflection in the line. Y = x. **find** the matrix which defines T.

(b)A transformation P maps points (1,3) and (-2,-3) on to points (2,4) and (-3,-11) respectively. Find the matrix of the transformation.

18. (a) **Draw** the graphs of y=2Cosx and $y = \dfrac{1}{Sin\ x}$ *for* $0^0 \le x \le 360$

(b) use your graph to solve
$$2Cos\ x = \dfrac{1}{Sin\ x} - 1$$

$$\dfrac{1}{Sin\ x} > 1$$

19. The diagram below shows a histogram representing marks obtained in a certain test.

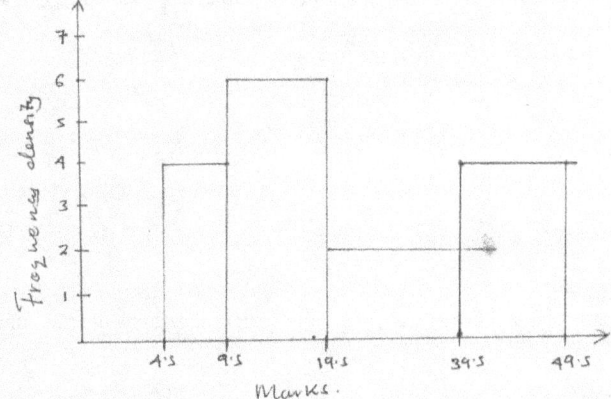

Develop a frequency distribution table for the data if the first class 5 -9 has a frequency of 8.

(a) **Estimate** the mean

15

(b) Calculate interquartile range.

20.

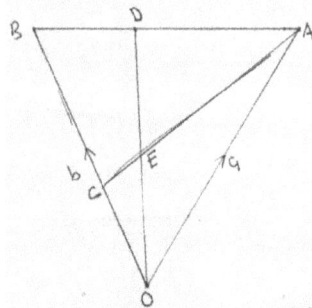

The figure above shows △OAB in which BD:DA = 1:2, OE: ED = 3:2 and c is the midpoint of OB.

Given that OA = a and OB = b express the following vectors in terms of

a) $\overrightarrow{AB}$

b) $\overrightarrow{OD}$

c) $\overrightarrow{AE}$

d) Show that points A, E and C lie on a straight line. Hence determine the ratio of CE: EA

21. (a) ABCD is a rectangle in which AB = 7.6 cm and AD = 5.2cm. Use a pair of compasses and a ruler only to construct rectangle ABCD, and construct the locus of a point P within the rectangle such that P is equidistant from CB and CD.

(b) Q is a variable point within the rectangle. ABCD drawn in (a) above such that $60^0 \leq$ AQB $\leq 90^0$. On the same diagram, construct and show the locus of point Q, by leaving unshaded, the region in which point Q lies.

22. An auto spare dealer sells two types of lubricant A and B in his shop. While purchasing type A cost Sh. 40 per 100ml tin and type B cost Sh. 60 per 100 ml tin. He decided to buy at lease 30 tins altogether of type A and B with Sh. 1500 available. He decides that at least one third of the tins should be of type B. He buys x tins of type A and y tins of type B.
(a)Write down three inequalities, which represent the above information.

(b)On a graph paper, **draw** a graph to show the three inequalities (a) above.

(c)**Determine** how many tins of each type that he should buy to maximize his profit if he makes a profit of sh. 10 of each type A tin and a profit of sh. 20 on each type B tin

(d)Calculate his maximum possible profit.

23. PQRSV is a right pyramid on a horizontal square base of side 10cm. the slant edges are all 8cm long. **Calculate**

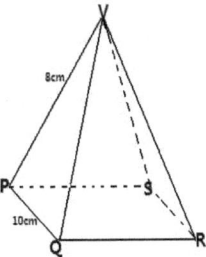

The height of the pyramid

The angle between
Line VP and the base PQRS

Line VP and line RS

Planes VPQ and the base PQRS.

Volume of the pyramid.

24. Two towns A and B lie on the same parallel of latitude 60^0N if the longitudes of A and B are 42^0W and 29^0E respectively.

Find the distance between A and B in nautical miles along the parallel of latitude.

Find the local time at A if at B is 1.00pm.

Find the shortest distance between A and B along the earth's surface in km.

$$\text{(Take } \pi = \frac{22}{7} \ \text{and} \ R = 6370 \ km \ )$$

If C is another town due south of A and 10010km away from A, **find** the coordinate C.

SOLUTIONS TO CHAPTER TWO

<table>
<tr><td>1.</td><td>

No   *Logs*

16.49^2 $2 \times 1.2172 \;\Rightarrow\; 2.4344$

$0.6329^{1/3}$ $\dfrac{1}{3} \times \bar{1}.8013$ $\bar{1}.9338 \;+$

                                  2.3682

438.2 $2.6417 \;-$

                                  $\bar{1}.7265$

0.53272 5.3272×10^{-1}

</td></tr>
<tr><td>2.</td><td>

Let $7^x = y$

$y^2 - 8y + 7 = 0$

$y^2 - 7y - y + 7 = 0$

$y(y-7) -,(y-7) = 0$ $(y-1)(y-7) = 0$

$y = 1$ or $y = 7$

$7^x = 1$ $x = 0$

$7^x = 7$ $x = 1$

</td></tr>
<tr><td>3.</td><td>

$\dfrac{dy}{dx} = 2x - 1$ $y = x^2 - x + c$

$5 = 1 - 1 + c$ $c = 5$

$y = x^2 - x + 5$

</td></tr>
<tr><td>4.</td><td>

$A\; P\left(1 - \dfrac{r}{100}\right)^n$

$250000 = 500000\left(1 - \dfrac{12}{100}\right)^n$

$0.5 \; 0.88^n$

   $\log 0.5 = n \log 0.88$

   $n = \dfrac{\log 0.5}{\log 0.88}$

$= 5.9459 \;\; yrs$

</td></tr>
<tr><td>5.</td><td>

$y = 5 - x$

$x(5-x) - 4 = 0$

$-x^2 + 5x - 4 = 0$

$-x^2 + 4x + x - 4 = 0$

$-x(x-4) + 1(x-4) = 0$

$x = 1$ $x = 4$

</td></tr>
</table>

	when x = 1 y = 4 when x = 4 y = 1 $\dfrac{2\,(0.1)}{4-0.1} = \dfrac{0.2}{3.9}$ $\dfrac{0.2}{3.9} \times \dfrac{10}{10}$ $= \dfrac{2}{39}$
6.	$27 \times 40 = 1080$ $\dfrac{1080}{30} + 1 = 37 \; beacons$
7.	$T_{13} = a + 12d = 1$ $\quad 7 + 12d = 1 \quad d = -\tfrac{1}{2}$ $\dfrac{n}{2}\{2a + (n-1)d\} = 0$ $14 + (n-1) \times -\tfrac{1}{2} = 0$ $-\tfrac{1}{2}n + \tfrac{29}{2} = 0$ $\quad n = 29$
8.	$\dfrac{22}{7}\,(10 \times 10 - 8 \times 8) \times 0.05 \; cm^3$ $22 \times 36 \times 0.05 \; cm^3$ $= 39.6 cm^3$
9.	$opp = \sqrt{5-4} = \sqrt{1} = 1$ (a) $Sin\ x = \dfrac{1}{\sqrt{5}}$ $= \dfrac{1}{\sqrt{5}} \times \dfrac{\sqrt{5}}{\sqrt{5}} = \dfrac{\sqrt{5}}{5}$ (b) $\tan(90 - x) = \dfrac{2}{1} = 2$

10.	2	3	5	7
2	22	32	52	72
3	23	33	53	73
5	25	35	55	75
7	27	37	57	77

(b) $\dfrac{4}{16}$ or $\dfrac{1}{4}$ or 0.25

---

11. (a) $a^5 - 5a^4b + 10a^3b^2 - 10a^2b^3 + 5ab^4 + b^5$

(b) $a = 2$

$b = 0.02$

$2^5 - 5 \times 2^4 \times 0.02 + 10 \times 2^3 \times 0.02^2$

$32 - 1.6 + 3.2 \times 10^{-2}$

$32 - 16 + 0.32 \Rightarrow 32.032 - 1.6$

$\qquad = 30.432$

---

12. $\angle ECA = 28^0 \ alt \ angles$

$\angle AEG = 28^0 \ alt \ seg.$

$(b) \ \angle CEG = 180 - (32 + 28) = 120^0$

$\therefore ABC = 60^0 \ opp. \angle \ S \ is \ cyclic \ quad.$

$\qquad add \ upto \ to \ 180^0$

13.

Frequency	4	12	18	11	4	1	$\sum f = 50$
X	28	33	38	43	48	53	
d= x-38	-10	-5	0	5	10	15	
$t = \dfrac{d}{5}$	-2	-1	0	1	2	3	
Ft	-8	-12	0	11	8	3	$\sum$ ft=2
ft2	16	12	0	11	16	9	$\sum$ ft2 = 64

$S.d = 5 \left(\sqrt{\dfrac{64}{50} - (\dfrac{2}{50})} \right)^2 = 5 \left(1.28 - 0.0016 \right)^{1/2}$

$\qquad 5 \times 1.1307 = 5.653$

---

14. $h^3 = \dfrac{c - x^2}{b}$

$c - x^2 = h^3 b$

$x^2 = c - h^3 b$

$x = \pm \sqrt{c - h^3 b}$

21

15.	$R = K + C\bar{I}$ $55 = K + 300C$ $58 = K + 330C$ $3 = 0 + 30C$ $30C = 3 \quad c = 0.1$ $55 = K + 0.1 \times 300$ $55 = K + 30 \quad k = 25$ $R = 25 + 0.1\bar{I}$
16.	x-section area $= \dfrac{22}{7} \times 3.5 \times 3.5 \Rightarrow 38.5cm^2$ $Volum \ / \sec \ = \dfrac{38.5}{10000} \times 1.5m^3 \ / \sec \ \Rightarrow$ $\quad 7.7 \quad 0.3 \quad 2$ $\dfrac{38.5}{10000} \times 1.5 \times 8 \times 3600 \, m^3 \ / \, day$ $\quad 25$ $\quad 5 \quad 1$ $No. \ of \ days \ \dfrac{4158}{7.7 \times 0.3 \times 2 \times 36} days$ $\qquad = 25 \ days$
17	(a) matrix transformation $\begin{pmatrix} 1 & 3 \\ 0 & 1 \end{pmatrix}$ Final matrix using unit square on graph $\begin{pmatrix} I^{11} & J^{11} \\ x_1 & x_2 \\ y_1 & y_2 \end{pmatrix} \begin{pmatrix} 0 & 1 \\ 1 & 3 \end{pmatrix}$

(b) $\begin{pmatrix} a & b \\ c & d \end{pmatrix}\begin{pmatrix} 1 & -2 \\ 3 & -3 \end{pmatrix} = \begin{pmatrix} 2 & -3 \\ 4 & -11 \end{pmatrix}$

$a + 3b = 2$

$\underline{-2a - 3b = -3}$

$-a \quad = -1$

$a = 1$

$1 + 3b = 2$

$b = \dfrac{1}{3}$

$c + 3d = 4$

$-2c - 3d = -11$

$-c = -7$

$7 + 3d = 4$

$d = -1$

matrix $\begin{pmatrix} 1 & \dfrac{1}{3} \\ 7 & -1 \end{pmatrix}$

18.	X	0	30	60	90	120	150	180	210	240	270	300	330	360
	2cosx	2	1.73	1.0	0	-1.0	-1.73	-2.0	-1.73	-1.0	0	1.0	1.73	2.0
	1/sin x	-1	∞	1		0.15	0	0.15	1	∞	-3	-2.15	-2	-2.15 -3∞

23

(b) (i) $12^0 \pm 2$

90^0

$0 \leq x < 90$

(ii) $90 < x \leq 180$

19.	(a) class	5 – 9	10 – 19	20 – 39	40 – 49
	Frequency	8	24	16	16

(b)

x	7	14.5	29.5	44.5	
f	8	24	16	16	
fx	56	348	472	712	$\sum fx = 1588$

$$\text{mean} = \frac{\sum fx}{\sum f} = \frac{1588}{64} = 24.8125$$

(c)

	5 – 9	10 – 19	20 – 39	40 – 49
f	8	24	16	16
cf	8	32	48	64

	$\dfrac{1}{4} \times 64 = 16$
	$Q = 9.5 + \dfrac{16 - 8}{24} \times 10$
	$9.5 + \dfrac{1}{3} \times 10 = 12.83$
	$Q = 19.5 + \dfrac{48 - 32}{16} \times 20$
	39.5
	Interquatile range
	$39.5 - 12.83$
	$= 26.67$
20.	(a) (i) $\overrightarrow{AB} = -\underset{\sim}{a} + \underset{\sim}{b}$
	(ii) $\overrightarrow{OD} = \dfrac{1}{3}\underset{\sim}{a} + \dfrac{2}{3}\underset{\sim}{b}$
	$\overrightarrow{OE} = \dfrac{3}{5}OD$
	$= \dfrac{1}{5}\underset{\sim}{a} + \dfrac{2}{5}\underset{\sim}{b}$
	(iii) $\overrightarrow{AE} = AO + OE$
	$= -\underset{\sim}{a} + \dfrac{1}{5}\underset{\sim}{a} + \dfrac{2}{5}\underset{\sim}{b}$
	$= -\dfrac{4}{5}\underset{\sim}{a} + \dfrac{2}{5}\underset{\sim}{b}$
	$\overrightarrow{AC} = -\underset{\sim}{a} + \dfrac{1}{2}\underset{\sim}{b}$
	$\overrightarrow{AE} = -\dfrac{4}{5}\underset{\sim}{a} + \dfrac{2}{5}\underset{\sim}{b}$
	(b) $AE = \dfrac{4}{5}AC$
	$AE \parallel AC$
	and A is common
	Hence A, E, C lie on one straight line
	$CE : EA$
	$1 : 4$

21	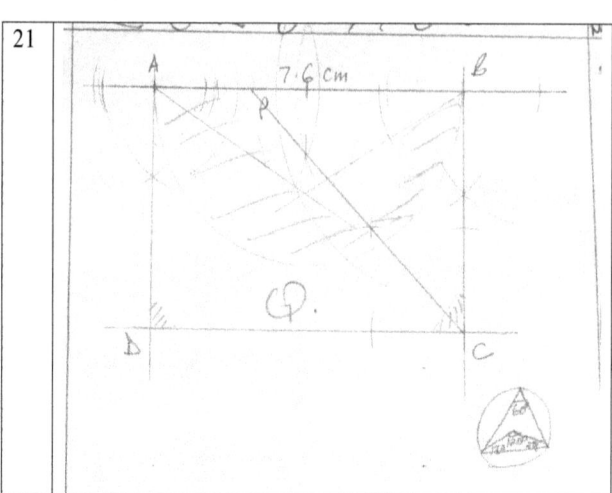
22	(i) $x + y \geq 30$ $40x + 60y \leq 1500$ (ii) $2x + 3y \leq 75$ (iii) $y \geq 10$ 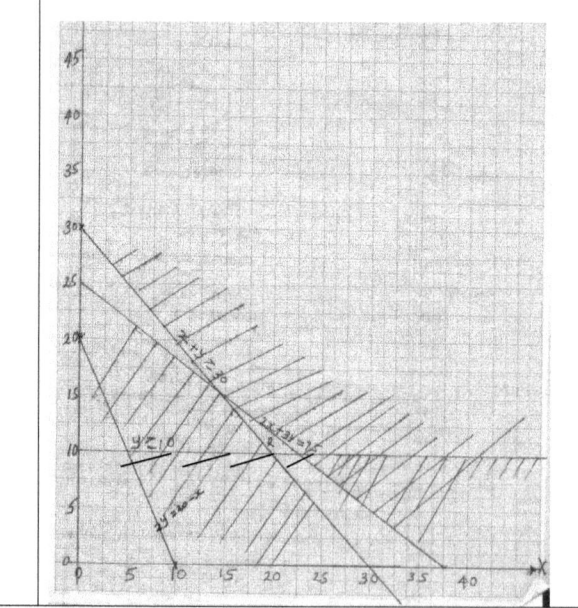

	$y \geq 10$ $10x + 20y = c$ $let \ c = 200$ $x = 21 \quad y = 11$ (iii) $21 \times 10 + 11 \times 20$ 210 + 220 430 $\quad Ksh$
23.	(a) $\quad Vo = \sqrt{8^2 - 7.071} = \sqrt{64 - 50}$ $\qquad = \sqrt{14} = 3.74$ (b) (i) $Cos \ \theta = \dfrac{7.071}{8} =$ $\qquad \theta = 27.89^0$ (ii) $Cos \ \alpha = \dfrac{5}{8} =$ $\qquad \alpha = 51.32^0$ (iii) $Tan \ \theta = \dfrac{3.74}{5}$ $\qquad \theta = 36.80^0$ (c) $\dfrac{1}{3} \times 100 \times 3.74$ $\qquad = 124.7 \ cm^3$
24.	(a) 71 x 60 cos 60 n.m $\qquad = 2130$ n.m (b) 71 x 4min = 284 min 4hrs 44 min $\quad$ 1300 – 4hrs 44min = 8.16 am $\quad Sin \ \dfrac{1}{2} \alpha = Sin \ 35.5 \ Cos60 \Rightarrow 0.29035$ $\qquad\qquad \alpha = 33.76^0$ (c) $\dfrac{33.76}{360} \times 2 \times \dfrac{22}{7} \times 6370 \ km \quad = \dfrac{33.76 \times 11 \times 91}{9}$ $\qquad\qquad\qquad = 3754.86 \ km$

$$\frac{\theta}{360} \times 2 \times \frac{22}{7} \times 6370 = 10010$$

(d) $\qquad \theta = \dfrac{10010 \times 7 \times 360}{2 \times 22 \times 6370} = 90$

$C\left(30^{0}S,\ 42^{0}W\right)$

CHAPTER THREE

1. Use Square roots, reciprocal and square tables to evaluate to 4 significant figures the expression.

$$(0.0546)^{\frac{1}{2}} + \left(\frac{2}{4.327}\right)^2$$

2. A line passes through the points A (2, 6) and B (4, -8). Find the equation of the perpendicular bisector of line AB.

3. Find the value of x which satisfies the equation.

$$16^{x^2} = 8^{4x-3}$$

4. A water tank has a capacity of 50 litres. A similar model tank has a capacity of 0.25litres. if the larger tank has a height of 100cm. calculate the height of the model tank.

5. Given that matrix $M = \begin{pmatrix} a & 0 \\ 5 & b \end{pmatrix}$

   (a) Determine M^2

   (b) If $M^2 = \begin{pmatrix} 1 & 0 \\ 0 & 1 \end{pmatrix}$, determine the possible pairs of values of a and b

6. Find the integral values that satisfy the inequality.

$$2x + 3 \geq 5x - 3 > -8$$

7. A tea blender buys two grades of tea at sh 60 and sh 80 per packet. Find the ratio in which he should mix them so that by selling the mixture at sh 90, a profit of 25% is realized.

8. A two digit number is such that when the digits are reversed the value of the number increases by 9. Three times the sum of its digits is less than the value of the number by 8. Find the number.

9.   From the roof of a house, a boy can see an avocado tree which is 20m away from the house. He measures the angle of elevation of the top of the tree as 21^0 and the angle of depression of the bottom of the tree as 31^0. Find the height of the avocado tree.

10.   Three people Makori, Ondieki and Mosomi contributed money to start a business. Makori contributed a quarter of the total amount and Ondieki two fifth of the remainder. Mosomi's contribution was one and half times that of Makori. They borrowed the rest of the money from the bank which was sh. 60,000 less than Mosomi's contribution. Find the total amount required to start the business.

11.   Draw line PQ = 7.0cm. Locate a point R on line PQ such that PR:RQ = 3:4

12.   From a survey carried out the following information was entered in a field book

                              Y
180 to N
To R 90   180
60 to M
    X
If XY is 360m and SM, RP and QN are the offsets. Determine the area of the field in metres.

13.   The figure below shows a net of a circular cone with a lid. Given that $\angle AOB = 150^0$ and
       OA = 14cm.

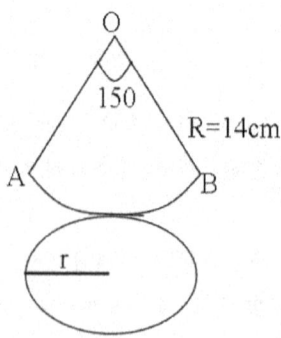

Determine

    (i)   The radius of the base of the cone

    (ii)   The total surface area of the cone

14.   The interior angle of a regular polygon is 9 times the exterior angle. How many sides does the polygon have?

15.   Solve for x in the equation

$$\log_3(x+23) - \log_2 16 + 1 = \log_3(9-x)$$

16.   The figure below represents a school field.

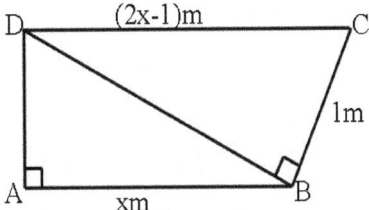

Find the length of AB given that $\angle BAD$ and $\angle CBD$ are right angles.

31

17. A plastic water tank has a shape as shown below, with a frustrum of a cone on top, a cylindrical body and a hemispherical bottom.

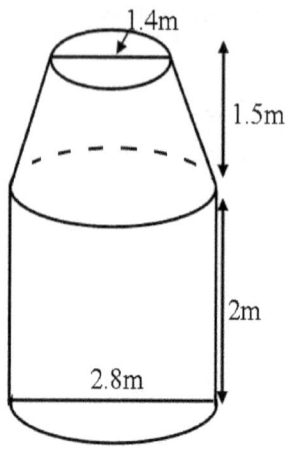

1.4m

1.5m

2m

2.8m

(a) Calculate

   (i) The volume of the tank in m³.

(b) A filler pipe takes 3 hours to fill a third of the tank. If the tank is already ¼ full, at what time will the filler pipe fill the tank if the pipe is opened at 9.00a.m.

(c) A particle falls in the tank. If its chances of being in any part of the tank are equally likely, find the probability of it being in the hemispherical part

18. In the figure below $\overrightarrow{OB} = b; OC = 3\overrightarrow{OB}$ and $OA = a$

   (a) Given that $OD = \dfrac{1}{3}OA$ and $AN = \dfrac{1}{2}AC$, CD and AB meet at M. Dtermine in

terms of a and b

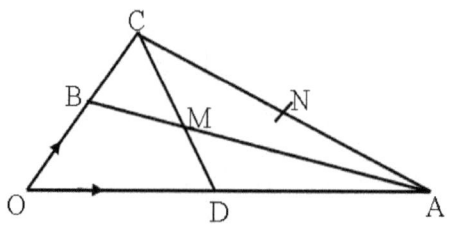

   (i) $\overrightarrow{AB}$

(ii) $\vec{CD}$

(b)   Given that $\vec{CM} = K\vec{CD}$ and $\vec{AM} = h\vec{AB}$. Determine the values of the scalars K and h.

(c)   Show that O,M and N are collinear.

19.   Construct a parallelogram ABCD in which AB = 8.5cm, AD = 6cm and angle BAD = 75^0. (Use a ruler and pair of compasses only in this question)

    (a)   Measure the length of AC

(b)   On the same diagram, construct

(i)   a perpendicular from B to line AD at M. Measure BM. Hence calculate the area of the parallelogram ABCD.

(ii)   The locus of a point x which moves such that it is equidistant from A and C

(iii)   The locus of point Y which moves such that angle BYD = 90^0

20.   The figure below shows two circles intersecting at C and D. The centres are A and B with radii 8cm and 6cm respectively. AB = 10cm.

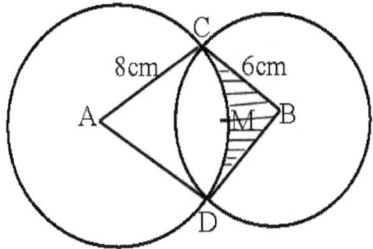

Determine
    (i)   Size of angle DAC

    (ii)   Size of angle DBC

    (iii)   Area of sector ACMD

    (iv)   Area of the shaded region

21.   In the figure below ABCDE is a pentagon inscribed in a circle. CX is a tangent to the circle at C and EDX is a straight line. Angle ADE = 34⁰, angle CAD = 42⁰, AB = BC and BC is parallel to AD. Giving reasons, determine

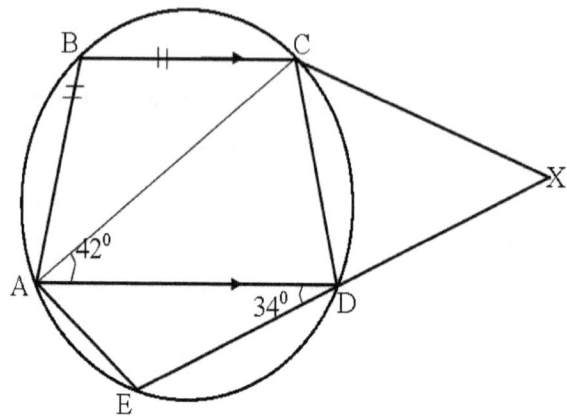

(a)   (i)   Angle ABC

(ii)   Angle ACD

(iii)   Angle EAD

(iv)   Angle CXD

(b)   Given that ED = 5cm, DX = 4cm, calculate the length of CX

22.   Complete the table below for the function
        $Y=2x^2 + 4x - 3$

X	-4	-3	-2	-1	0	1	2
$2x^2$	32		8			2	
4x-3			-11				5
Y			-3			3	13

(a)   On the grid provided draw the graph of the function $Y=2x^2 + 4x - 3$ for $-4 \le x \le 2$ and use your graph to estimate the roots of the equation.

(i) $2x^2+4x-3=0$

(ii) $2x^2-x-5=0$

34

(b)   Determine the equation of the tangent to the curve $y=2x^2+4x-3$ at the point $x=1$

23.   (a)   Every time an insect jumps forward the distance covered is half of the previous jump.

If the insect initially jumped 6.4m. Calculate

        (i)   The length of 7th jump

        (ii)   The total distance covered after the 7th jump

(b)   An arithmetic progression has the first term as a and common difference as d

        (i)   Write down the third, ninth and twenty fifth term of the progression

        (ii)   The arithmetic progression above is such that it is increasing and the third, ninth and twenty fifth terms form the first three consecutive terms e.g. a geometric progression. If the sum of the seventh term and twice the sixth term of the A.P is 78.Calculte the first term and the common difference of A.P.

24.   (a)   Draw the graphs of $y = \sin 3x$ and $y = \cos (x + 30^0)$. On the same axis for $-180^0 \le x \le 180^0$. Take an interval of 30^0.

(b)   Use the graph to solve

(i) $\cos(x+30^0) - \sin 3x = 0$

(ii) $3 \cos (x + 30^0) - 2 = 0$

SOLUTIONS TO CHAPTER THREE

1 $(5.46 \times 10^{-2})^{½} + 4 \,(0.0534)$

$(5.46 \times 10^{-2})^{½} + 4 \dfrac{1}{18.72}$

$2.337 \times 10^{-1} + 4 \left(\dfrac{1}{1.872 \times 10^{1}} \right)$

$0.2337 + 4 \,(0.534 \times 10^{-1})$

$= 0.2337 + 4 \,(0.00534)$

$= 0.2337 + 0.2136$

$= 0.4473$

---

2 Mid point $= \left(\dfrac{2+4}{2}, \dfrac{6+-8}{2} \right)$

$\qquad = (3, -1)$

A $(2, 6)$ B $(4, -8)$

Grad of line AB $= \dfrac{-8-6}{4-2} = \dfrac{-14}{2} = -7$

Grad of bisector $= \dfrac{1}{7}$

Eqtn of $\perp$ an $\dfrac{y--1}{x-3} = \dfrac{1}{7}$

$\qquad\qquad 7y + 7 = x - 3$

$\qquad\qquad 7y - x = -10$

$\qquad\qquad y = \dfrac{x-10}{7}$

---

3 $2^{4x^2} = 2^{3(4x-3)}$

$4x^2 = 3\,(4x - 3)$

$4x^2 = 12x - 9$

$4x^2 = 12x + 9 = 0$

$(2x - 3)\,(2x - 3) = 0$

$x = 1.5$

4	$V.S.F = \dfrac{50}{0.25} = 200$
	$L.S.F = \sqrt[3]{200}\ = 5.848$
	height of smaller tank $= \dfrac{100}{5.848} = 17.10$ cm

5	$M^2 = \begin{pmatrix} a & 0 \\ 5 & b \end{pmatrix}\begin{pmatrix} a & 0 \\ 5 & b \end{pmatrix}$
	$M^2 = \begin{pmatrix} a^2 & 0 \\ 5a+5b & b^2 \end{pmatrix}$
	$\begin{pmatrix} a^2 & 0 \\ 5a+5b & b^2 \end{pmatrix} = \begin{pmatrix} 1 & 0 \\ 0 & 1 \end{pmatrix}$
	$a^2 = 1, a = \sqrt{1}\ = \pm\,1$
	when a =1, b = -1
	and when a = -1, b = 1

6	$2x + 3 \geq 5x - 3$ $\qquad$ $5x - 3 > -8$
	$6 \geq 3x$ $\qquad\qquad$ $5x > -5$
	$2 \geq x$ $\qquad\qquad$ $x \geq -1$
	$\qquad\qquad\qquad\qquad$ $x > -1$
	$\qquad\qquad -1 < x \leq 2$
	Integral values 0, 1, 2

7	Profit = 100 % + 25 % = 125 %
	125% = 90
	100% = ?
	$\dfrac{90 \times 100}{125}$ = sh. 72
	Let the ratio of grade 1 = x
	$\qquad\qquad\qquad\quad$ 2 = y
	$\dfrac{60x + 80y}{x + y} = 72$
	60x + 80y = 72x + 72y
	8y = 12x
	$\dfrac{8}{12} = \dfrac{x}{y}$
	$\dfrac{x}{y} = \dfrac{2}{3}$
	x : y = 2 : 3

8	Let the number be $10x + y$ Reversing $(10y + x) - (10x + y) = 9$ $\qquad 9y - 9x = 9$ $\qquad y - x = 1$ $(10x + y) - 3(x + y) = 8$ $\qquad 7x - 2y = 8$ $\quad 7x - 2y = 8$ $\underline{-2x + 2y = 2}$ $\qquad x = 2$ $\qquad y = 3$ The number $10x + y = 23$

9	

$$\tan 21 = \frac{x}{20}$$

$x = 20 \tan 21$

$$\tan 31 = \frac{y}{20}$$

$y = 20 \tan 31$

Height $= x + y = 20 \tan 21° + 20 \tan 31°$
$$\qquad\qquad = 7.677 + 12.017$$
$$\qquad\qquad = 19.694$$

10	Makori = ¼x

Ondieki $= \dfrac{2}{5} \times \dfrac{3}{4}x = \dfrac{3}{10}x$

Mosomi $= \dfrac{3}{2} \times \dfrac{1}{4}x = \dfrac{3}{2} \times \dfrac{1}{4}x = \dfrac{3}{8}x$

Money borrowed $= x - \left(\dfrac{1}{4}x + \dfrac{3}{10}x + \dfrac{3}{8}x\right) = \dfrac{3}{40}x$

$$\dfrac{3}{8}x - \dfrac{3}{40}x = 60,000$$

| 11 | x = 200,000 |
| | Total money = Kshs. 200,000 |

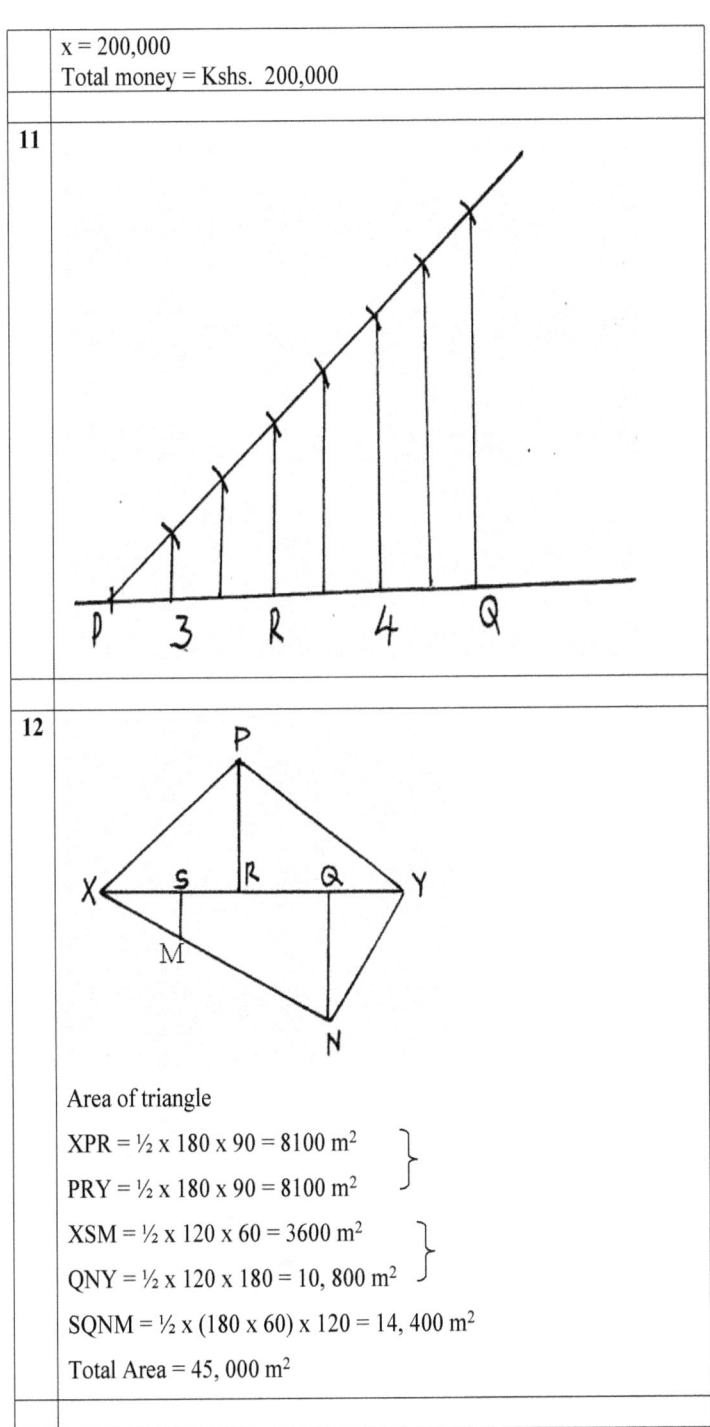

| 12 | |

Area of triangle

XPR = ½ x 180 x 90 = 8100 m²

PRY = ½ x 180 x 90 = 8100 m²

XSM = ½ x 120 x 60 = 3600 m²

QNY = ½ x 120 x 180 = 10, 800 m²

SQNM = ½ x (180 x 60) x 120 = 14, 400 m²

Total Area = 45, 000 m²

13	$\dfrac{22}{7} \times 2r = \dfrac{22}{7} \times \dfrac{150}{360} \times 2 \times 14$
	$r = \dfrac{35}{6} = 5\dfrac{5}{6}$
	$S.A = \dfrac{22}{7} \times 14 \times \dfrac{35}{6} + \dfrac{22}{7} \times \left(\dfrac{35}{6}\right)^2$
	$= \dfrac{22}{7} \times \dfrac{35}{6}\left(14 + \dfrac{35}{6}\right)$
	$= 363.611$
14	Ext. angle + int. angle = 180°
	$x + 9x = 180$
	$10x = 180 \Rightarrow x = 18$
	$n = \dfrac{360}{x} = \dfrac{360}{18} = 20$ sides
15	$\text{Log}_3\,(x + 23) - 4\log_2 2 + 1 = \text{Log}_3\,(9 - 3)$
	$\text{Log}_3\,(x + 23) - 4 + 1 = \log_3\,(9 - x)$
	$\text{Log}_3\,(x + 23) - 3 = \log_3\,(9 \text{ - } x)$
	$\text{Log}_3\,(x + 23) - \log_3 3^3 = \log_3\,(9 - x)$
	$\text{Log}_3\left(\dfrac{x + 23}{27}\right) = \log_3\,(9 - x)$
	$x + 23 = 27\,(9 - x)$
	$x + 23 = 243 - 27x$
	$28x = 220$
	$x = 7.857$ or $7\,{}^6/_7$
16	$DB^2 = x^2 + 1$
	$DB^2 + BC^2 = DC^2$
	$x^2 + 1 + 1 = (2x \text{ - } 1)^2$
	$x^2 + 2 = 4x^2 - 4x + 1$
	$3x^2 - 4x - 1 = 0$
	$x = \dfrac{4 \pm \sqrt{16 + 12}}{6}$

x = 1.55 or – 0.215

Ignore – 0.215 – length never –ve.

AB = XM = 1.55m

17

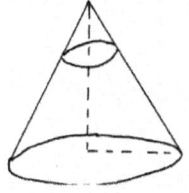

L.s.f $= \dfrac{2.8}{1.4} = 2$

$\dfrac{H}{h} = 2$

$\dfrac{h + 1.5}{h} = 2$

h = 1.5

a) Volume = volume of hemisphere + cylinder + frustrum

$= \dfrac{1}{3}\pi\,(1.4^2 \text{ x } 3 - 0.7^2 \text{ x } 1.5)$ +

$\dfrac{22}{7} \times 1.4 \times 1.4 \times 2 + \dfrac{2}{3} \times 1.4^3 \text{ x } ^{22}/_7$

$= \dfrac{1}{3} \times \dfrac{22}{7}(5.88 - 0.735) + 12.32 + 5.749$

$= 5.39 + 12.32 + 5.749$

$= 23.46\text{m}^3$

b) to fill full tank $= \dfrac{1}{1/3} = 9\text{hrs.}$

in hr $= \dfrac{1}{9}$ of tank is filled.

to fill $\dfrac{3}{4}$ of tank $= \dfrac{3}{4} \times \dfrac{9}{1} = 6\dfrac{3}{4}$ hrs.

$= 6$ hrs 45 mins

Time to fill tank = 9.00 a.m + 6 hrs 45 mins

= 3.45 p.m

c) $\dfrac{volume\ of\ hemisphere}{volume\ of\ tank} = \dfrac{5.749}{23.46}$

$= 0.245$

18

a)

$AB = -\underset{\sim}{a} + \underset{\sim}{b}$

$\quad = \underset{\sim}{b} - \underset{\sim}{a}$

$CD = -3\underset{\sim}{b} + \dfrac{1}{3}\underset{\sim}{a}$

$\quad = \dfrac{1}{3}\underset{\sim}{a} - 3\underset{\sim}{b}$

b)

$\quad CM = k\left(\dfrac{1}{3}\underset{\sim}{a} - 3\underset{\sim}{b}\right) = \frac{1}{3}ka - 3kb$

$\quad AM = h\,AB$

$\qquad = h\underset{\sim}{b} - h\underset{\sim}{a}$

$\quad OM = \overrightarrow{OC} + \overrightarrow{CM} = 3\underset{\sim}{b} + \dfrac{1}{3}k\underset{\sim}{a} - 3k\underset{\sim}{b}$

$\qquad\qquad\qquad\qquad = (3 - 3k)\underset{\sim}{b} + \dfrac{1}{3}k\underset{\sim}{a}$

$\quad OM = \overrightarrow{OA} + AM = \underset{\sim}{a} + h\underset{\sim}{b} - h\underset{\sim}{a}$

$\qquad\qquad\qquad\qquad = (1 - h)\underset{\sim}{a} + h\underset{\sim}{b}.$

$\quad (3 - 3k)\underset{\sim}{b} + \dfrac{1}{3}k\underset{\sim}{a} = (1 - h)\underset{\sim}{a} + h\underset{\sim}{b}$

$\qquad\qquad 3 - 3k = h\ldots\ldots\ldots\ldots\text{ (i)}$

$\qquad\qquad\quad \dfrac{1}{3}k = 1 - h\ldots\ldots\ldots\text{ (ii)}$

$\qquad\qquad\quad \dfrac{1}{3}k = 1 - (3 - 3k)$

$\qquad\qquad\qquad k = \dfrac{3}{4}$

$\qquad\qquad\qquad h = 3 - 3\left(\dfrac{3}{4}\right)$

$\qquad\qquad\qquad h = \dfrac{3}{4}$

c)

$\overrightarrow{ON} = \underset{\sim}{a} + \dfrac{1}{2}(3\underset{\sim}{b} - \underset{\sim}{a})$

$$= \frac{1}{2}(a + 3b)$$

$$OM = \left(1 - \frac{3}{4}\right)a + \frac{3}{4}b$$

$$= \frac{1}{4}(a + 3b)$$

$$OM = \frac{1}{4}(2\,0N)$$

$$OM = \frac{1}{20}\,ON$$

OM is parallel to ON
O is the common point
Hence O, M & N are collinear

19

b) (i)   BM = 8.3 cm ± 0.1
        Area = base x height
             = 6cm x 8.3cm
             = 49.8 cm²

20

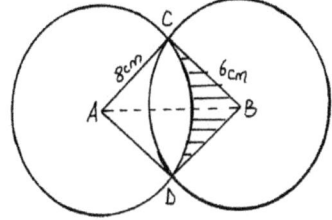

i)

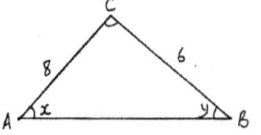

$$6^2 = 8^2 + 10^2 - 2 \times 8 \times 10 \cos x$$

a)
$$Cos\ x = \frac{128}{160} = 0.8$$

$$x = 36.87°$$

$$\angle DAC = 2x = 73.74°$$

(ii) $\angle DBC$

$$8^2 - 6^2 + 10^2 - 2 \times 6 \times 10 \cos y$$

$$Cos\ y = \frac{72}{120} = 0.6$$

$y = 53.13°$

$\angle DBC = 2y = 106.26°$

(iii) Area of sector ACD.

$$= \frac{73.74°}{360°} \times \frac{22}{7} \times 8 \times 8$$

$$= 41.83 \text{cm}^2$$

(iv) Area of quadrilateral ABCD

$$= \frac{1}{2} \times 8 \times 8 \sin 73.74 + \frac{1}{2} \times 6 \times 6 \sin 106.6°$$

$$= 30.72 + 17.28 = 48 \text{cm}^2$$

Area of shaded region

$$= 48 \text{cm}^2 - 41.83 \text{cm}^2$$

$$= 6.17 \text{cm}^2$$

21	

(i) $\angle$ BCA = 42° alternating $\angle$ to CAD

$\quad \angle$ BAC = 42° = $\angle$ BCA $\angle$s subtended by same chord.

$\quad \angle ABC = 180° - (42° + 42°) = 96°$

(ii) $\angle BAD + \angle BCD = 180°$ (opp. $\angle$s of cyclic quad)

$\quad \angle BCD = 96°$

$\quad \angle ACD = 96° - 42° = 54°$

iii) $\angle EAD = 180° - (34° + 126°)$ ($\angle$s in a cyclic quad)

$\quad = 180° - 160°$

$\quad = 20°$

iv) $\angle XDC = 180° - (34 + 34) = 52°$ $\angle$s in a straight line

$\quad \angle DCX = \angle CAD = 42°$ alternate

$\quad \angle CXD = 180° - (42° + 52°) = 86°$

(v) $Cx^2 = XD \times xE$

$\quad = 4 \times 9$

$\quad = 36$

$\quad CX = 6\text{cm}$

22	

x	-4	-3	-2	-1	0	1	2
$2x^2$	32	18	8	2	0	2	8
4x -3	-19	-15	11	-7	-3	1	5
y	13	3	-3	-5	-3	3	13

a) (i)

x = - 2.55 and x = 0.55 ± 0.1

(ii)

$$2x^2 + 4x - 3 = y$$
$$\underline{- 2x^2 + x - 5 = 0}$$
$$3x + 2 = y$$

x	0	2
y	2	8

x = - 1.8 and x = 1.4

b)

$$y = 2x^2 + 4x - 3$$

$$\frac{dy}{dx} = 4x + 4$$

gradient = 4 + 4 = 8

$$\frac{y - 3}{x - 1} = 8$$

y - 3 = 8x - 8   ⟹ y = 8x - 5

---

23 a) i) a = 6.4

r = ½

Tn = ar^{n-1}

$$= 6.4 \ x \left(\frac{1}{2}\right) = 0.1$$

ii) $Sn = \dfrac{a\left(1 - r^n\right)}{1 - r}$

$$= \frac{6.4\left(1-\left(\frac{1}{2}\right)^6\right)}{1-\frac{1}{2}}$$

$$= 12.8\left(1-0.015625\right)$$

$$= 12.6M$$

b) (i) Tn = a + (n - 1)d

T3 = a + 2d

T9 = a + 8d

T25 = a + 24d

(ii) a + 2d, a + 8d, a + 24d

$$\text{G.P.} = \frac{a+8d}{a+2d} = \frac{a+24d}{a+8d}$$

(a + 8d)(a + 8d) = (a + 2d)(a + 24d)

$a^2 + 16ad + 64d^2 = a^2 + 24ad + 48d^2$

5a = 8d..............(ii)

a + 6d + 2(a + 5d) = 78

3a + 16d = 78

3a + 10a = 78

13a = 78

a = 6

d = $^5/_8$a = $^5/_8$ x 6

$$d = 3\frac{3}{4}$$

24

CHAPTER FOUR

1. Use logarithms tables to evaluate

$$\sqrt[3]{\frac{36.72 \times (0.46)^2}{185.4}}$$

2. Given that $\dfrac{2\sqrt{2}}{1+\sqrt{3}} - \dfrac{\sqrt{3}}{1-\sqrt{3}} = a + b\sqrt{c}$. Find the values of a, b and c.

3. Given that the equation of a curve is $y = (2x + 2)(x^2 - 3)$

   (i) Find the function of the gradient of the curve and its value when $x = \dfrac{3}{2}$

   (ii) Determine the equation of the normal to the curve at the point (-2, 3)

4. A quantity f varies partly as t and partly as the square root of t. When t = 4, f = 22 and when t = 9, f = 42. Write the equation connecting f and t.

5. (a) Expand $(2 + 2x)^5$ up to the forth term.
   (b) Hence find the value of $(2.02)^5$ correct to 3 decimal place

6. Find the distance between the centre A of a circle whose equation is $2x^2 + 2y^2 + 6x + 10y + 7 = 0$ and the point B (-4, 1)

7. Machine A can do a piece of work in 8hours while machine B can do the same piece of work in 10 hours. The two machines were set to do the work together but after 2 ½ hours B broke down leaving A alone to complete the rest of the work. How long did it take machine A to do the remaining work.

8. T is a transformation represented by the matrix $\begin{pmatrix} 5x & 2 \\ -3 & x \end{pmatrix}$ under T, a square of area 10cm²

is mapped onto a square of area 110cm². Find the value of x

9. Given that $2\cos(2x-30^0) = -\dfrac{6}{5}$ find x where $180^0 \le x \le 360^0$

10. A cooker is valued at Ksh. 8000 if it appreciated by 10% in the first year, 12% in the second year and then 8% per annum in the subsequent years, determine its value at the end of 4 years.

11. Make A the subject of the formula

$$T = \frac{2m}{n}\sqrt{\frac{L-A}{3K}}$$

12. The sides of a triangle were measured to 1 decimal place as 6.5cm, 7.4cm and 8.2cm respectively. Calculate the percentage error in its perimeter

13. Given that (5m – 2n) : (2m – n) = 7:5. find the ratio m:n

14. A line L_1 is perpendicular to the line 2y + 3x = 6. Determine the acute angle made by the line L_1 and the x – axis

15. A bus and a matatu starts from Nairobi to Kisii via Narok at the same time making a distance of 280km. The matatu averages 20km/h faster than the bus and reaches there 1 hour 36minutes earlier. Determine the speed of the bus.

16. In the figure below the diameter AB of the circle is parallel to DC. DCE is a straight line and angle BCE = 63⁰. Calculate the angle DBC.

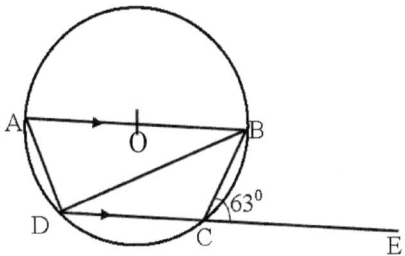

17. Mr Ondati is a salaried civil servant. He earns a basic monthly salary of sh. 20,640, a house allowance of sh. 6,800p.m and medical allowance of sh. 2800p.m. He claims a family relief of sh. 400p.m. He pays sh. 300 per month and 2% of his salary towards water bills and NHIF respectively.

Calculate his net monthly salary in Ksh using the tax rates shown in the table below.

k£ p.a	Rate in Ksh per £
1 – 1980	2
1981 – 3960	3
3961 – 5940	5
5941 – 7920	7
7921 – 9900	9
9901 and over	10

18.   In the figure below EFGHIJKL is a square based frustrum whose dimensions are as shown. The perpendicular height of the frustrum is 9cm. Given that EF = FG = GH = HE = 10cm and JK = KL = IL = IJ = 4cm.

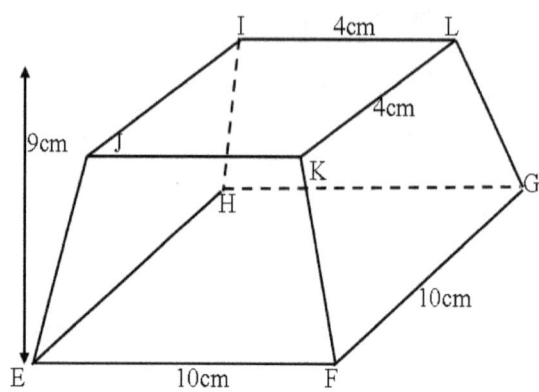

(a)   Calculate
       (i)   The altitude of the pyramid.
       (ii)   The angle between the line FK and the base EFGH
       (iii)   The angle between line LG and EF
(b)   The volume of the frustrum

19.   Complete the table for the function $y = (x + 1)(x - 2)^2$ for $-2 \le x \le 3$
       (a)

x	-2	-1	0	1	2	3
y		0		4		4

       Draw the curve of function $y = (x + 1)(x - 2)^2$ in the domain $-2 \le x \le 3$ on a grid.
(b)   Using the mid-ordinate rule and strips of equal width of 0.5 estimate the area enclosed by the curve and the x – axis.
(c)   Find the exact area in (c) above
(d)   Calculate the percentage error in the area (b) above

20. The position of two towns A and B on the earths surface are (36^0N, $49E^0$) and (36^0 N, 131^0 W) respectively.

    (a)   Find the difference in longitude between town A and town B

    (b)   Given that the radius of the earth is 6370, calculate the distance between town A and town B

(i)   In nm

(ii)   In kilometers

    (c)   Another town C is 840km East of town B on the same latitude of town A and B. Find the longitude of town C

21. In a certain mathematical relationship, the values of A and B are observed to satisfy the relationship $B = CA + KA^2$ where C and K are constants. Below is the table of values of A and B.

A	1	2	3	4	5	6
B	3.2	6.75	10.8	15.1	20	25.2

22. The table below shows the scores of Mathematics of a particular class in a certain school.

Marks	1-10	11-20	21-30	31-40	41-50	51-60	61-70	71-80	81-90	91-100
Frequency	4	6	7	5	x	9	3	5	2	7

By taking an assumed mean to be 45.5 marks, the value of actual mean is 49.5 marks.

    (a)   Determine the value of x

    (b)   Calculate the standard deviation

    (c)   If 30 students passed the test. Calculate the pass mark

23. The probability of Mary, Esther and John coming to school late on Monday are $\frac{1}{4}$, $\frac{2}{5}$ and $\frac{1}{3}$ respectively.

    (a)   Draw a tree diagram to represent the information.

    (b)   Calculate the probability that

        (i)   All the three girls are late

        (ii)   All except Esther are late

        (iii)   At most two girls are late

24. A carpenter makes two types of chairs for Keroka Secondary School. To make type A chair it requires 6 man – hours where as a type B requires 4 man – hours. The cost of material for type A is sh 120 and that for type B is sh100. The profit on type A is sh 80 and profit on type B is sh 60. The carpenter has to abide by the following conditions

    (i)   A contract to supply 15 of type A and 10 of type B per week has to be fulfilled

    (ii)   Only 300 man – hours are available in each week.

    (iii)   Total weekly cost of material for all chairs should not exceed sh 6000

        If type A and type B chairs were x and y per week respectively.

    (a)   Write down the inequalities satisfying these conditions

    (b)   Represent this information on a grid and show the region by shading out the unwanted.

    (c)   What values of x and y will give maximum profit. Determine this maximum profit.

SOLUITIONS TO CHAPTER FOUR

1

No	Log
36.72	1.5649
$(0.46)^2 \Rightarrow 2\left(\overline{1}.6628\right)$	$\overline{1}.3256 +$
	0.8905
	2.2682 −
	$\overline{2}.6223$
	$\overline{2}.6223$
	3
	$\overline{3}+1.6223$
	3
3.474×10^{-1}	$\overline{1}.5408$
$= 0.3474$	

2

$$= \frac{2\sqrt{3}}{1+\sqrt{3}} - \frac{\sqrt{3}}{1-\sqrt{3}}$$

$$= \frac{2\sqrt{3}\left(1-\sqrt{3}\right) - \sqrt{3}\left(1+\sqrt{3}\right)}{1-\sqrt{3}+\sqrt{3}-3}$$

$$= \frac{\sqrt{3}-9}{-2}$$

$$= \frac{-9}{-2} + \frac{\sqrt{3}}{-2}$$

$$= \frac{9}{2} + \frac{\sqrt{3}}{-2} = a+b\sqrt{c}$$

$$a = \frac{9}{2}, b = -\frac{1}{2}, c = 3$$

3

$$y = 2x^3 - 6x + 2x^2 - 6$$

$$\frac{dy}{dx} = 6x^2 - 6 + 4x$$

Gradient when $x = \dfrac{3}{2}$

$$= 6 \times \left(\frac{3}{2}\right)^2 - 6 + 4 \times \frac{3}{2}$$

$$= \frac{54}{4} - 6 + 6$$

$$= 13.5$$

Grad of normal $= -\dfrac{2}{27}$

$$\frac{y-3}{x+2} = -\frac{2}{27}$$

$27(y - 3) = -2(x + 2)$

$27y - 81 = -2x - 4$

$27y + 2x = 77$

---

4 $f = kt + m\sqrt{t}$

when t = 4, f = 22

$22 = 4k + 2m$ ....(i)

When t = 9, f = 42

$42 = 9k + 3m$ ....(ii)

$42 = 9k + 3m$

$14 = 3k + m$

$$\begin{array}{r} 14 = 3k + m \\ \underline{11 = 2k + m} \\ 3 = k \end{array}$$

$m = 11 - 2 \times 3 = 5$

Equation $f = 3t + 5\sqrt{t}$

---

5 $(2 + 2x)^5 = 2^5 + 5(2)^4(2x) + 10(2)^3(2x)^2 + 10(2)^2$

$(2x)^3 + 5(2)(2x)^4 + \ldots\ldots$

$= 32 + 160x + 320x^2 + 320x^3 + 160x^4 + \ldots\ldots$

$(2 + 2x)^5 = (2.02)^5$

$2x = 2.02 - 2$

$2x = 0.02$

$x = 0.01$

$32 + 160(0.01) + 320(0.01)^2 + 320(0.01)^3 +$

$160(0.01)^4 + \ldots\ldots\ldots\ldots$

$= 32 + 1.6 + 0.032 + 0.00032 + \ldots\ldots\ldots$

$= 33.632$

---

6 $\quad 2x^2 + 2y^2 + 6x - 10y + 7 = 0$

$x^2 + y^2 + 3x - 5y + 3.5 = 0$

Centre **A** (- 1.5, 2.5), **B** (-4, 1)

$= \sqrt{\left(-4--1.5\right)^2 + \left(1 - 2.5\right)^2}$

$= \sqrt{\left(-2.5\right)^2 + \left(-1.5\right)^2}$

$= \sqrt{6.25 + 2.25}$

$= \sqrt{8.5}$

$= 2.9155$ units

---

7 Fraction done by **A** and **B** in $\dfrac{5}{2}$ hrs

$= \dfrac{5}{2}\left(\dfrac{1}{8} + \dfrac{1}{10}\right)$

$= \dfrac{9}{16}$

Remaining work $= \dfrac{16}{16} - \dfrac{9}{16} = \dfrac{7}{16}$

Time taken by **A** alone

$\dfrac{7}{16} \div \dfrac{1}{8} = 3\ \tfrac{1}{2}$ hrs

---

8 A.S.F = det

$5x^2 + 6 = \dfrac{110}{10}$

$5x^2 + 6 - 11 = 0$

$5x^2 = 5$

$x = \pm 1$

---

9 $\quad 2\cos(2x - 30) = -\dfrac{6}{5}$

$\cos(2x - 30) = -0.6$

Cos −ve in 3rd quad

$2x - 30 = 53.13$

$2x - 30 = 233.13$

$2x = 263.10$

$x = 131.57°$

$2x - 30° = 360 + 233.13°$

$2x = 623.13$

$x = 311.57$

---

10 1ST year

$\dfrac{110}{100} \times 8000 = \text{sh } 8,\ 800$

2nd year

$\dfrac{112}{100} \times 8800 = \text{sh } 9,\ 856$

$A = P \left(1 + \dfrac{R}{100} \right)^n$

$= 9856\ (1.08)^2$

$A = \text{sh } 11,\ 496$

---

11 $\dfrac{nT}{2m} = \sqrt{\dfrac{L - A}{3k}}$

$\dfrac{n^2 T^2}{4m^2} = \dfrac{L - A}{3K}$

$4m^2\ (L - A) = 3k\ (n^2\ T^2)$

$4Lm^2 - 4Am^2 = 3kn^2T^2$

$A = \dfrac{4LM^2 - 3kn^2T^2}{4m^2}$

---

12 Absolute error $= 0.05$

Actual perimeter $= 6.5 + 7.4 + 8.2$

$= 22.1$ cm

Max. perimeter $= 6.55 + 7.45 + 8.25$

$$= 22.25 \text{ cm}$$

Min. perimeter $= 6.45 + 7.35 + 8.15$

$$= 21.95$$

% error $= \dfrac{1}{2}\left(\dfrac{22.25 - 21.95}{21.1}\right) \times 100\%$

$$= \dfrac{0.3}{44.2} \times 100\%$$

$$= 0.679\,\%$$

13

$\dfrac{5m - 2n}{2m - n} = \dfrac{7}{5}$

$5(5m - 2n) = 7\,(2m - n)$

$25m - 10n = 14m - 7n$

$11m = 3n$

$\dfrac{m}{n} = \dfrac{3}{11}$

$m : n = 3 : 11$

14

$2y + 3x = 6$

$y = -\dfrac{3}{2}x + 3$

gradient $= -\dfrac{3}{2}$

gradient $= \dfrac{2}{3}$

$\text{Tan}^{-1}\dfrac{2}{3} = 33.69°$

15

Bus time $= \dfrac{280}{x}$, matatu $= \dfrac{280}{x + 20}$

$\dfrac{280}{x} - 1.6 = \dfrac{280}{x + 20}$

$280\,(x + 20) - 1.6x\,(x + 20) = 280x$

$16x^2 + 320x - 56000 = 0$

$$x = \frac{-320 \pm \sqrt{320 + 4 \times 16 \times 56000}}{2 \times 16}$$

$$= \frac{-320 \pm 1920}{32}$$

$$= 50$$

Bus speed = 50km/h

---

16 $\angle DCB = 180 - 63 = 117\,°$

$\angle DAB + \angle DCB = 180°$ ($\angle$s in cyclic quad)

$\angle DAB = 180 - 117$

$\qquad = 63°$

$\angle ADB = 90°$

$\angle DBC = 180 - (63 + 90°)$

$\qquad = 27°$

---

17 Taxable income per year

$$= K£ \frac{(20640 + 6800 + 2800) \times 12}{20}$$

$= K£\ 18144$

1st slab	1980 x 2	sh. 3960
2nd slab	1980 x 3	sh. 5940
3rd slab	1980 x 5	sh. 9900
4th slab	1980 x 7	sh. 13860
5th slab	1980 x 9	sh. 17820
Remaining slab 8244 x 10		sh. 82440

Total gross tax $= sh.\,13920\,p.a$

Gross tax per month $= \dfrac{133920}{12}$

$\qquad = sh.\ 11, 160\ p.m$

Less relief $=$ sh $(11, 160 - 400) = 10\ 760$ p.m

Total deduction $= \left(10760 + 300 + \dfrac{2}{100} \times 20640\right)$

$\qquad = sh.\ 11472.80$

Net pay $=$ sh$(30240 - 11472.80)$

$\qquad = sh.\ 18767.20$ p.m

18 a)

$$\frac{h+9}{h} = \frac{10}{4}$$

4h + 36 = 10h

6h = 36

h = 6

Height = 6 + 9 = 15cm

b)

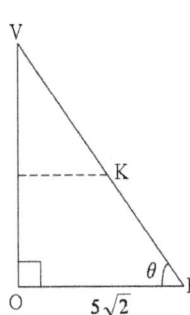

$OF = \frac{1}{2}\sqrt{10^2 + 10^2}$

$OF = 5\sqrt{2}$

$\text{Tan } \theta = \frac{15}{5\sqrt{2}} = 2.121$

$\angle VFO = 64.76°$

c) Translate EF to HG

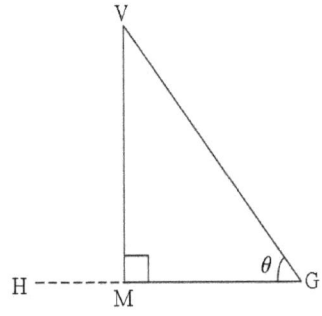

$\angle VGH = \angle VGM$

$VM = \sqrt{15^2 + 5^2}$

$= 15.814 \text{ cm}$

$\text{Tan } \theta = \frac{15.814}{5} = 3.162$

$\angle VGM = 72.45°$

d)

$\frac{1}{3} \times 10 \times 10 \times 15\,cm^3 - \frac{1}{3} \times 4 \times 4 \times 6\ cm^3$

$= 468\ cm^3$

19 a)

$y = x^3 - 3x^2 + 4$

x	-2	-1	0	1	2	3	2.5
y	-16	0	4	2	0	4	0.875

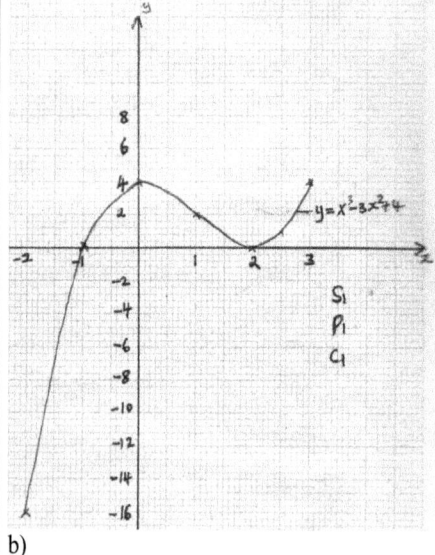

b)

x	-0.75	-0.25	0.25	0.75	1.25	1.75
y	1.5	3.4	3.8	2.6	1.4	0.3

Area $= 0.5\ (1.5 + 3.4 + 3.8 + 2.6 + 1.4 + 0.3)$
$= 6.5$ sq. units

c)

Exact area $= \int_1^2 \left(x^3 - 3x^2 + 4 \right) dx$

$= \left[\dfrac{x^4}{4} - x^3 + 4x \right]_{-1}^{2}$

$= \left(\dfrac{16}{4} - 8 + 8 \right) - \left(\dfrac{1}{4} + 1 - 4 \right)$

$$= 6\frac{3}{4} \text{ sq. units}$$

d)
error = 6.75 – 6.5 = 0.25

% error = $\dfrac{0.25}{6.75} \times 100 = 3.704\%$

---

20 a)
131° + 49° = 180°

b) (i)
60 x ∝ Cosθ

$= 60 \times 180° \times \cos 36°$

$= 60 \times 180 \times 0.8090$

$= 8737.38 \text{nm}$

(ii)

$$\frac{\alpha}{360°} \times 2\pi R \cos\theta$$

$$\frac{180}{360°} \times 2 \times \frac{22}{7} \times 6370 \cos 36°$$

$= 16192.103 \text{ km}$

c)
$$\frac{\alpha}{360} \times 2 \times \frac{22}{7} \times 6370 \cos 36° = 840 km$$

$$\alpha = \frac{840 \times 360°}{628 \times 6370 \cos 36°}$$

$\alpha = 9.34°$

Longitude of town **C**

131° - 9.34°

= 121.66°

---

21 B = CA + KA²

$\dfrac{B}{A} = C + KA$

Plot $\dfrac{B}{A}$ against A

C= y – intercept

K = gradient

A	1	2	3	4	5	6
B	3.2	6.75	10.8	15.1	20	25.2
$\dfrac{B}{A}$	3.2	3.375	3.6	3.775	4.0	4.2

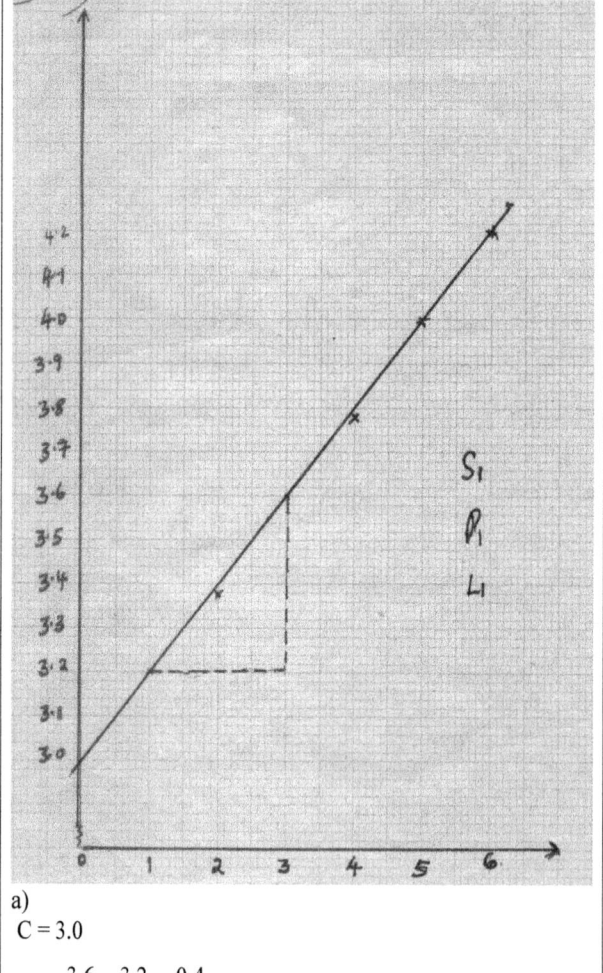

a)

C = 3.0

$$K = \frac{3.6 - 3.2}{3 - 1} = \frac{0.4}{2} = 0.2$$

b)

$B = 3A + 0.2A^2$ or $B = 3A + \dfrac{1}{5}A^2$

c)
$B = 3 \times 7 + 0.2 \times 7^2$

$= 21 + 9.8$

$= 30.8$

22

Marks	f	x	d x-A	fd	d²	fd²
1-10	4	5.5	-40	-160	1600	6400
11-20	6	15.5	-30	-180	900	5400
21-20	7	25.5	-20	-140	400	2800
31-40	5	35.5	-10	-50	100	500
41-50	x	45.5	0	0	0	0
51-60	9	55.5	10	90	100	900
61-70	3	65.5	20	60	400	1200
71-80	5	75.5	30	150	900	4500
81-90	2	85.5	40	80	1600	3200
91-100	7	95.5	50	350	2500	17500
	Σfd=			Σfd=200		Σfd²=**42,400**

a) (i)

$45.5 + \dfrac{200}{x+48} = 49.5$

$\dfrac{200}{x+48} = 4$

$200 = 4x + 192$

$x = 2$

(ii)

$S = \sqrt{\dfrac{\sum fd^2}{\sum f} - \left(\dfrac{\sum fd}{\sum f}\right)^2}$

$\quad = \sqrt{\dfrac{42\,400}{50} - \left(\dfrac{200}{50}\right)^2}$

$\quad = \sqrt{848 - 16}$

$\quad = \sqrt{832}$

$\quad = 28.82$

(iii)

Both values

$$\left[30.5 + \frac{3}{5} \times 10\right]$$

$= 36.5$

23 a)

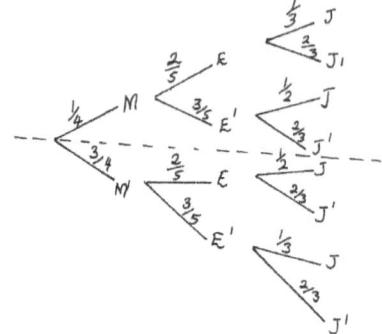

b) (i) P (all late) $= \dfrac{1}{4} \times \dfrac{2}{5} \times \dfrac{1}{3}$

$$= \frac{2}{60}$$

$$= \frac{1}{30}$$

(ii) P (all except E) $= \dfrac{1}{4} \times \dfrac{3}{5} \times \dfrac{1}{3}$

$$= \frac{3}{60}$$

$$= \frac{1}{20}$$

(iii) P (at least one late)

$$\left(\frac{1}{4} \times \frac{2}{5} \times \frac{1}{3}\right) + \left(\frac{1}{4} \times \frac{2}{5} \times \frac{2}{3}\right) + \left(\frac{1}{4} \times \frac{3}{5} \times \frac{1}{3}\right) +$$

$$\left(\frac{3}{4} \times \frac{2}{5} \times \frac{1}{5}\right) + \left(\frac{3}{4} \times \frac{2}{3} \times \frac{1}{3}\right)$$

$$= \frac{3}{30} + \frac{4}{20} + \frac{2}{10} + \frac{1}{5}$$

$$= \frac{21}{30}$$

$$= \frac{7}{10}$$

OR

1 – (P none is late)

$1 - \left(\dfrac{3}{4} \times \dfrac{3}{5} \times \dfrac{2}{3} \right)$

$= \dfrac{7}{10}$

(iv)

P (at most 2)

1 – (P all late)

$= 1 - \left(\dfrac{1}{4} \times \dfrac{2}{5} \times \dfrac{1}{3} \right)$

$= 1 - \dfrac{1}{30}$

$= \dfrac{29}{30}$

---

24 a)

$6x + 4y \leq 300$

$x \geq 15$

$y \geq 10$

$120x + 100y \leq 6000$

b)

c)

point (15, 42)

x = 15, y = 42

max. profit

(15 x 80) + (42 x 60)

1200 + 2520

Sh. 3720

CHAPTER FIVE

1. Without using a calculator or mathematical tables, evaluate;

$$\frac{5}{6} - \frac{1}{3} of \frac{27}{20} \div 2$$

2. Find the equation of the perpendicular bisector of the line AB where the coordinates of A and B are (-3, 2) and (6, 4) respectively.

3. Three bells P, Q and R are programmed to ring after an interval of 15 minutes, 25 minutes and 50   minutes respectively. If they all rang together at 8.45 a.m, when will they next ring together again.

4. Simplify the expression $\dfrac{x+4}{x-4} - \dfrac{3x+12}{x^2 - 16}$

5. O is the centre of the circle below and AB is parallel to DC. Angle ACD = 70^0 and angle ACB = 10^0.

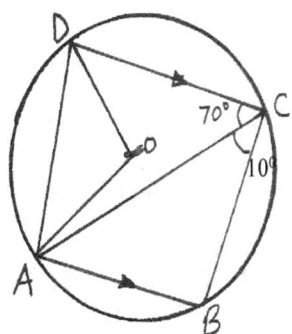

    Calculate angles

ABC

OAD

6. A prism of length 20cm is represented by the diagram below whose cross section is an equilateral   triangle of side 7cm.

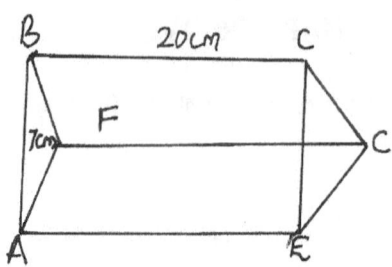

(a) Draw a sketch net of the prism and label it correctly.

(b) Calculate

    (i) The area of the triangular end

    (ii) The total surface area of the prism

    (iii) The volume of the prism.

7. Solve the following inequalities and represent the solutions on a single number line:

$$2 - 2x < 4$$
$$-6 - 3x \geq -15$$

8. Solve for n in

$$\left(\frac{1}{49}\right)^{n} \times (343)^{-1} = 7$$

9. Solve the simultaneous equations

$$\log_4 (2x + 3y) = 2$$
$$\log_2 (4x - y) = 2$$

10. In 2007 parliamentary election, only 55% of the voters in a constituency of 85,000 cast their

votes. Of the votes cast, A received 48%, B received 32% and C received the remainder. How many votes did C receive.

11. If each interior angle of a regular polygon is 150^0, how many sides does the polygon have?

12. The expression $1 - \frac{x}{2}$ is taken as an approximation for $\sqrt{1 - x}$. Calculate the percentage

error in   doing so when $x = \frac{7}{16}$.

13. (a) In the diagram below find the length of EC if BC = 12 cm.

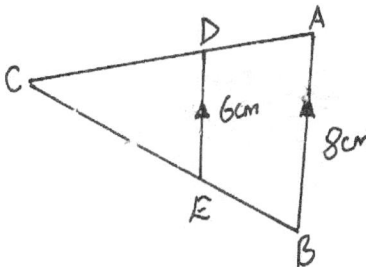

(b) Given that the area of triangle DCE is 27 cm², find the area of ABECD.

14. Point T is the midpoint of a straight line AB. Given that the position vectors of A and T

are $-\underset{\sim}{i}+\underset{\sim}{j}-\underset{\sim}{k}$ and $3\underset{\sim}{i}+4\underset{\sim}{j}$ respectively, find the position vector of B in terms of $\underset{\sim}{i}, \underset{\sim}{j}$ and $\underset{\sim}{k}$

15. Given the coordinates of P, Q and R as (2, -1), (3, 4) and (6, 2) respectively, find the

coordinates of P¹, Q¹ and R¹ the images of P, Q and R under a transformation represented by

the matrix. $\begin{pmatrix} -1 & 2 \\ 3 & 1 \end{pmatrix}$.

16. Given that θ is an acute angle and $\sin\theta = \dfrac{2\sqrt{3}}{5}$, find without using calculators or

mathematical table, $\tan(90-x)^0$.

17. Two circles with centres O_1 and O_2, have radii 7cm and 6cm respectively. The two circles

intersect at P and Q and the length of the common chord PQ is 10cm.

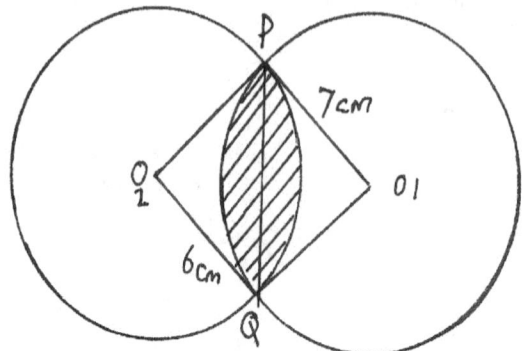

Calculate the area of the shaded region in the above diagram to 4 significant figures.

18.   On some day, Mr. Makori bought some oranges worth ksh. 45. On another day of the same week,   Mrs Makori spent the same amount of money but bought the oranges at a discount of 75 cents per   orange.

   (a)   If Mr. Makori bought an orange at sh x, write down a simplified expression for the total   number of oranges bought by the two in the week.

   (b)   If Mrs. Makori bought 2 oranges more than her husband, find how much each spent on an   orange.

   (c)   Find the number of oranges bought for the family that week.

19.   A cone is made by cutting off a sector as shown below from a circle and gluing the straight edges of the sector. The cone formed has slant height 14cm and circular base of perimeter 11cm   (take $\pi = \dfrac{22}{7}$)

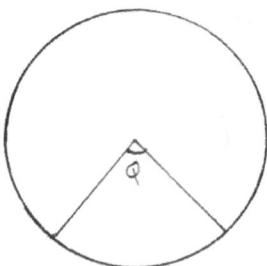

(a)   Calculate the value of θ.
   (b)   The radius of the cone's circular base
   (c)   The height of the cone.
   (d)   The cone is cut uniformly on a horizontal plane 1cm below the apex. Calculate the slant height of the frustum so formed correct to 2 decimal places.

70

20. (a)   Draw the graph of $y = 2 + 3x - x^2$ in the range of $-3 \le x \le 6$ on the grid provided.
  b)   From your graph:-
        (i)   Find the value of x if $x^2 - 4x = 0$
        (ii)   Determine the value of x for which y is the greatest.
        (iii) Determine the range of values of x for which y is positive.

21. A Kenyan businesswoman wants to pay a company she owes US$ 100,000 in the united
  states of   America. The woman can either pay through her account in Kenya or through her
  account in the united kingdom.
        (a)   If the exchange rate is;
                1 US Dollar = 28.74 Kenya shillings
                1 Sterling Pound = 1.79 US Dollars
                1 Sterling Pound = 50.80 Kenya shillings,
              Which method is cheaper and by how much? Give your answer in Kenya
  shillings.
        (b)   Three years ago, Joseph was three times older than Agnes. In two years time the
  sum of their ages will be 75. Determine their present ages.
        (c)   By use of reciprocals, evaluate the following and give answer to 3 decimal places
                $$\frac{3}{0.0416} + \frac{5}{49.27}$$

22. A circular path of width 14 metrres surrounds a field of diameter 70 metres. The path is
  to be   carpeted and the field to have a concrete slab with an exception of four rectangular holes
  each   measuring 4 metres by 3 metres. A contractor estimated the cost of carpeting the path at
  sh. 300   per square metre and the cost of putting the concrete slab at sh. 400 per square
  metre. He then   made a quotation which was 15% more than the total estimate. After
  completing the job, he   realizes that 20% of the quotation was not spent.
        (a)   How much money was not spent?
        (b)   Find the actual cost of the cotract
23. A transformation represents by the matrix $\begin{pmatrix} 2 & 0 \\ 0 & 2 \end{pmatrix}$ maps A(1, 3), B(3, 3) and C(2, 1) onto

  A^1B^1 and C^1 respectively.
        (a)   (i) On the grid provided, draw the triangle ABC and its image $A^1B^1C^1$ on the
  same axes.

              (ii) Hence or otherwise determine the area of the triangle $A^1B^1C^1$

        (b)   Another transformation represented by the matrix $\begin{pmatrix} 0 & 1 \\ 1 & 0 \end{pmatrix}$ maps $A^1B^1C^1$ onto

  $A^{11}B^{11}C^{11}$.
        (i)   Plot triangle $A^{11}B^{11}C^{11}$ on the same axes.

        (ii)   Describe the transformation represented by the matrix $\begin{pmatrix} 0 & 1 \\ 1 & 0 \end{pmatrix}$

(c)   Determine the matrix of the single transformation which maps $A^{11}B^{11}C^{11}$ onto ABC.

24.   The table below shows the height of tree in Manga forest in metres.

Height (m)	0-9	10-19	20-29	30-39	40-49	50-59	60-69
No of trees	50	35	30	32	16	10	7

Using an assumed mean of 34.5, calculate:-
(a)   The mean
(b)   The median
(c)   The standard deviation

SOLUTIONS TO CHAPTER FIVE

No.	Working
1.	$\dfrac{5}{6}-\left(\dfrac{1}{3}\times\dfrac{27}{20}\right)\div 2$
	$\dfrac{5}{6}-\left(\dfrac{9}{20}\times\dfrac{1}{2}\right)$
	$\dfrac{5}{6}-\dfrac{9}{40}$
	$\dfrac{100-27}{120}$
	$\dfrac{73}{120}$
2.	Grad. Of AB $=\dfrac{2-4}{-3-6}=\dfrac{-2}{-9}=\dfrac{2}{9}$
	Mid point of AB $=\left(\dfrac{-3+6}{2},\dfrac{2+4}{2}\right)$
	$\qquad = (1.5, 3)$
	$\left(\dfrac{y-3}{x-1.5}\right)\dfrac{2}{9}=-1$
	$\dfrac{2y-6}{9x-13.5}=-1$
	$2y-6=-1(9x-13.5)$
	$2y-6=-9x+13.5$
	$2y=-9x+19.5$
	$y=-\dfrac{9}{2}x+\dfrac{19.5}{2}$
	$or\ y=-4\dfrac{1}{2}x+9\dfrac{3}{4}$
3.	$15 = 3\times 5$
	$25 = 5\times 5 = 5^2$
	$50 = 2\times 5\times 5 = 2\times 5^2$
	$L.C.M = 2\times 3\times 5^2$
	$\qquad = 150$

4.	$\dfrac{x+4}{x-4} - \dfrac{3(x+4)}{(x-4)(x+4)}$
	$\dfrac{x+4}{x-4} - \dfrac{3}{x-4}$
	$\dfrac{x+4-3}{x-4}$
	$\dfrac{x+1}{x-4}$

5. (i)	∠ABC=180⁰-(10+70)⁰
	= 180-80⁰
	= 100⁰
(ii)	∠OAD = $\dfrac{180^0-(2\times70^0)}{2}$
	= $\dfrac{40^0}{2}$
	= 20⁰

6. (b)	
	(i) Area Δ end =

74

$\dfrac{1}{2} \times 7 \times 7 \sin 60^0$

$= 21.21762239$

$\cong 21.22 cm^2 \; 2 d.p$

(ii) Total surface area=

$3 \times 20 \times 7 + 2 \times \dfrac{1}{2} \times 7 \times 7 \sin 60^0$

$= 420 + 49 \sin 60^0$

$= 462.432448$

$\cong 462.44 cm^2 \quad 2 d.p$

(iii) Volume

$= 20 \times \dfrac{1}{2} \times 7 \times 7 \sin 60^0$

$490 \sin 60^0$

$= 424.3524479$

$\cong 424.35 cm^2 \; 2 d.p$

7.	$-2x < 2$
	$x > -1 \, or -1 < x$
	$-3x \geq -9$
	$x \leq 3 \qquad\qquad i.e \qquad -1 \leq x \leq 3$

8.	$7^{-2n} \times 7^{-3} = 7^1$
	$7^{-2n-3} = 7^1$
	$-2n - 3 = 1$
	$-2n = 4$
	$n = -2$

9.	$2x + 3y = 4^2$
	$4x - y = 2^2$
	$\Leftrightarrow 2x + 3y = 16$.........$(i)$ $\times 1$
	$4x - y = 4$............(ii) $\times 3$
	$2x + 3y = 16$
	$12x - 3y = 12$
	$14x + 0 = 28$
	$14x = 28$
	$x = 2$
	$2(2) + 3y = 16$
	$3y = 12$
	$y = 4$
10	No. who voted $= \dfrac{55}{100} \times 85000 = 46,750$
	Votes received by C $= \dfrac{20}{100} \times 46,750$
	$= 9350$
11	$(2n - 4)90 = 150n$
	$180n - 360 = 150n$
	$30n = 360$
	$n = 12$

12	$1 - \dfrac{1}{2}\left(\dfrac{7}{16}\right)$
	$= 1 - \dfrac{7}{32}$
	$= \dfrac{32 - 7}{32}$
	$= \dfrac{25}{32}$
	$\sqrt{1-x} = \sqrt{1 - \dfrac{7}{16}}$
	$= \sqrt{\dfrac{9}{16}} \qquad = \dfrac{3}{4}$
	$error = \dfrac{25}{32} - \dfrac{3}{4}$
	$= \dfrac{25 - 24}{32}$
	$= \dfrac{1}{32}$
	$\%error = \dfrac{\frac{1}{32}}{\frac{3}{4}} \times 100$
	$= \dfrac{1}{32} \times \dfrac{4}{3} \times 100$
	$= \dfrac{100}{24}$
	$= 4\dfrac{1}{6}\% \, or \, 4.167\%$
13	(a) $\triangle ABC$ and $\triangle DEC$ are similar
	$\dfrac{DE}{AB} = \dfrac{EC}{BC}$
	$\dfrac{6}{8} = \dfrac{EC}{12}$
	$EC = \dfrac{6}{8} \times 12$
	$= 9cm$
	(b) linear scale factor $= 6:8$
	$= 3:4$
	Area scale factor $= 9:16$

	Area of $\triangle ABC$ $= \dfrac{16}{9} \times area\, of\, \triangle DCE$ $= \dfrac{16}{9} \times 27$ $= 48 cm^2$ $\therefore area\, of\, ABECD = 48 - 27$ $= 21 cm^2$
14.	 $OB = OA + 2\,AT$ $= OA + 2\left(-OA + OT\right)$ $= OA - 2\,OA + 2\,OT$ $= -OA + 2\,OT$ $= -\left(i - j + k\right) + 2\left(2i + 1\dfrac{1}{2}k\right)$ $= -i + j - k + 4i + 3k$ $\therefore OB = 3i + j + 2k$
15	$\qquad P \quad Q \quad R$ $\begin{pmatrix} -1 & 2 \\ 3 & 1 \end{pmatrix}\begin{pmatrix} 2 & 3 & 6 \\ -1 & 4 & 4 \end{pmatrix}$ $\qquad P^1 \ Q^1 \ R^1$ $= \begin{pmatrix} -4 & 5 & -2 \\ 5 & 13 & 20 \end{pmatrix}$ $P^1(-4,20)\, Q^1(5,13)\, R^1(-2,20)$

16	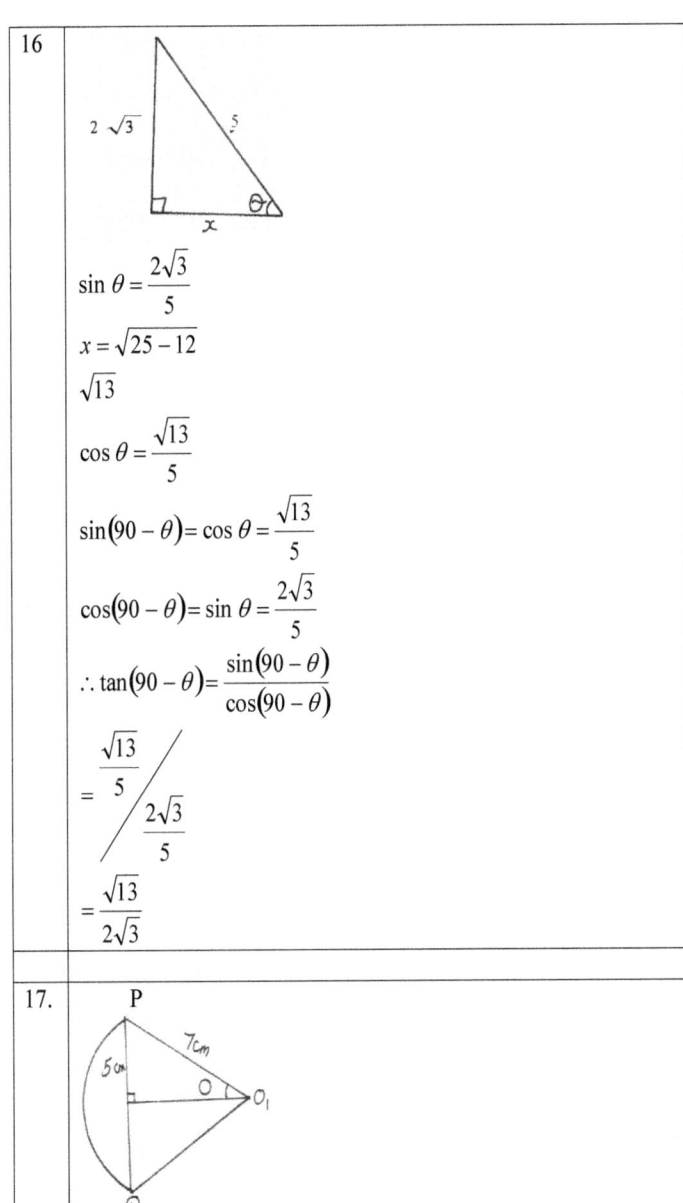 $$\sin \theta = \frac{2\sqrt{3}}{5}$$ $$x = \sqrt{25 - 12}$$ $$\sqrt{13}$$ $$\cos \theta = \frac{\sqrt{13}}{5}$$ $$\sin(90 - \theta) = \cos \theta = \frac{\sqrt{13}}{5}$$ $$\cos(90 - \theta) = \sin \theta = \frac{2\sqrt{3}}{5}$$ $$\therefore \tan(90 - \theta) = \frac{\sin(90 - \theta)}{\cos(90 - \theta)}$$ $$= \frac{\sqrt{13}}{5} \bigg/ \frac{2\sqrt{3}}{5}$$ $$= \frac{\sqrt{13}}{2\sqrt{3}}$$

17.

P

7cm

5cm

O

O₁

Q

$\sin Q = \dfrac{5}{7} = 0.7143$

$Q = 45.59^{0}$

$\angle PO_1Q = 91.17^{0}$

Area of sector PO_1Q

$= \dfrac{91.17^{0}}{360^{0}} \times \pi \times 7^{2}\, cm^{2}$

$= 38.98 cm^{2}$

Arrea of triangle PO_1Q

$= \dfrac{1}{2} \times 7^{2} \times \sin 91.17$

$= 24.49$

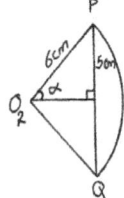

$\sin \alpha = \dfrac{5}{6} = 0.8333$

$\alpha = 56.439^{0}$

$\angle PO_2Q = 112.9^{0}$

Arrea of sector PO_2Q

$= \dfrac{112.9}{360} \times \pi \times 6^{2}$

$= 35.47 cm^{2}$

area of triangle PO_2Q

$= \dfrac{1}{2} \times 6^{2} \times \sin 112.9^{2}$

$= 16.58 cm^{2}$

area of shadded region

$= (38.98 - 24.49 + 35.47 - 16.58) cm^{2}$

$= 33.38 cm^{2}$

18.	$\dfrac{45}{x}+\dfrac{45}{x-0.75}$
	$\dfrac{45(x-0.75)+45x}{x(x-0.75)}$

$$\dfrac{45x-33.75+45x}{x^2-0.75x}=\dfrac{90x-33.75}{x^2-0.75x}$$

(b)

$$\dfrac{45}{x}+2=\dfrac{45}{x-0.75}$$

$$\dfrac{45+2x}{x}=\dfrac{45}{x-0.75}$$

$$(45+2x)(x-0.75)=45x$$

$$45x-33.75+2x^2-1.5x=45x$$

$$2x^2-1.5x-33.75=0$$

$$x=\dfrac{-(-1.5)\pm\sqrt{(1.5)^2-4(2)(-33.75)}}{2(2)}$$

$$=\dfrac{1.5\pm\sqrt{2.25+270}}{4}$$

$$=\dfrac{1.5\pm\sqrt{272.25}}{4}$$

$$=\dfrac{1.5\pm16.5}{4}$$

$$x=\dfrac{1.5\pm16.5}{4}\;or\;x=\dfrac{1.5-16.5}{4}$$

$$=\dfrac{18}{4}$$

$$x=4.5\;\;or\;x=-3.75$$

Therefore makori spent sh.4.50 per orange

Mrs. Makori spent sh. $(4.50-0.75)$

$$=\text{sh. }3.75$$

(c) no. of oranges for family that week

$$=\dfrac{45}{4.50}+\dfrac{45}{3.45}$$

$$=10+12$$

$$=22$$

19.	(a)

$$\frac{\theta}{360} \times 14 \times 2 \times \pi = 11$$

$$\theta = 45^0$$

$(b)\, 2\pi r = 11$

$$r = \frac{11}{2\pi}$$

$$= 1.75cm$$

(c)

$$h = \sqrt{14^2 - 1.75^2}$$

$$= \sqrt{196 - 3.0625}$$

$$= \sqrt{192.9375}$$

$$= 13.89cm$$

(d)

$$\frac{x}{14} = \frac{1}{13.89}$$

$$x = 14 \times \frac{1}{13.89}$$

$$= 1.0079$$

$$= 1.01cm$$

20.	(a)										
	x	-3	-2	-1	0	1	2	3	4	5	6
	y	-16	-8	-2	2	4	4	2	-2	-8	-16
	(b) (i)										

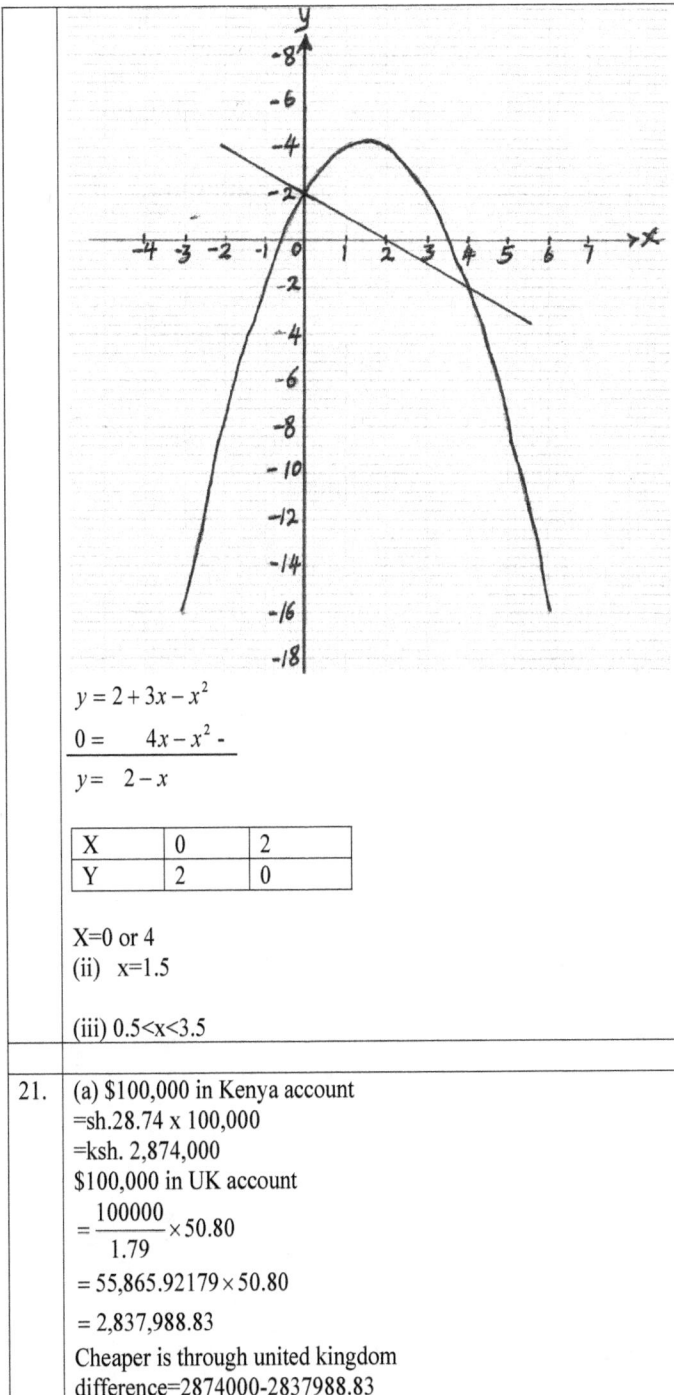

$y = 2 + 3x - x^2$

$0 = \quad 4x - x^2 -$

$y = \quad 2 - x$

X	0	2
Y	2	0

X=0 or 4

(ii) x=1.5

(iii) 0.5<x<3.5

21.	(a) $100,000 in Kenya account =sh.28.74 x 100,000 =ksh. 2,874,000 $100,000 in UK account $=\dfrac{100000}{1.79} \times 50.80$ $= 55,865.92179 \times 50.80$ $= 2,837,988.83$ Cheaper is through united kingdom difference=2874000-2837988.83

=sh.36.01

(b) Let Agnes age now be x

Three years ago Agnes x-3

Joseph three years ago 3(x-3)

Joseph now 3(x-3)+3

In two years time Agnes (x-3) +2=x+5

Joseph 3(x-3)+3+2

= 3x-4

(x-3)+2+3(x-3)+5=75

x-3+2+3x-9+5=75

4x = 80

X = 20

Agnes age now is 20

Josephs age now is 54

(c)

$$3\left(\frac{1}{0.416}\right)+5\left(\frac{1}{49.27}\right)$$

$$3\left(\frac{1}{4.16\times10^{-2}}\right)+5\left(\frac{1}{4.927\times10}\right)$$

$$3\left(\frac{10^2}{4.16}\right)+5\left(\frac{1}{4.927}\times\frac{1}{10}\right)$$

$$300\left(\frac{1}{4.16}\right)+\frac{5}{10}\left(\frac{1}{4.927}\right)$$

$$300(0.2404)+\frac{1}{2}(0.2030)$$

72.12 + 0.1015

72.222   (3d.p)

---

22. | (a)

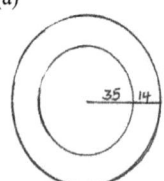

Area of path

$\pi\times49^2 - \pi\times35^2$

$= \pi\left(49^2 - 35^2\right)$

$= \left(2401-1225\right)\pi$

$= 3694.5m^2$

Area of field

84

	$= \pi \times 35^2 - (4 \times 3) \times 4$
	$= 3800.5m^2$
	Total estimate
	=3694.5x300+3800.5x400
	=1,108,350+1,520,200
	=sh.2,628,550
	Quotation
	$= \dfrac{15}{100} \times 2628550 + 2628550$
	$= 394282.5 + 2628550$
	$= sh.3,022,832.5$
	Money not spent
	$= \dfrac{20}{100} \times 3022832.5$
	$= sh.604,566.5$
	(b) Actual cost of contact
	$= sh.\ 3022832.5{-}604,566.5$
	$= 2,418,266$
23	(a)

$$\begin{array}{cc} & A \quad B \quad C \qquad A' \quad B' \quad C' \end{array}$$

$$\begin{pmatrix} 2 & 0 \\ 0 & 2 \end{pmatrix}\begin{pmatrix} 1 & 3 & 2 \\ 3 & 3 & 1 \end{pmatrix} = \begin{pmatrix} 2 & 6 & 4 \\ 6 & 6 & 2 \end{pmatrix}$$

$A'(2,6), \quad B'(6,6), \quad C'(4,2)$

(i)

$(ii) \, area \, of \, \Delta A'B'C' = \dfrac{1}{2} \times 4 \times 4 = 8 Sq \, units$

$$\begin{array}{cc} & A' \quad B' \quad C' \qquad A'' \quad B'' \quad C'' \end{array}$$

$(b) \begin{pmatrix} 0 & 1 \\ 1 & 0 \end{pmatrix}\begin{pmatrix} 2 & 6 & 4 \\ 6 & 6 & 2 \end{pmatrix} = \begin{pmatrix} 6 & 6 & 2 \\ 2 & 6 & 4 \end{pmatrix}$

$A''(6,2), \quad B''(6,6), \quad C''(2,4)$

(i)

$(ii) \, it \, is \, a \, reflection \, in \, the \, line \, y = x$

$$\begin{array}{cc} & A'' \quad B'' \quad C'' \qquad A \quad B \quad C \end{array}$$

$(c) \begin{pmatrix} a & b \\ c & d \end{pmatrix}\begin{pmatrix} 6 & 6 & 2 \\ 2 & 6 & 4 \end{pmatrix} = \begin{pmatrix} 1 & 3 & 2 \\ 3 & 3 & 1 \end{pmatrix}$

$\Rightarrow 6a + 2b = 1$

$\quad 6a + 6b = 3$

$\qquad -4b = -2$

$\qquad b = \dfrac{1}{2}$

$6a + \dfrac{1}{2} \times 2 = 1$

$6a = 0,$

$a = 0$

$\Rightarrow 6c + 2d = 3$

$\quad 6c + 6d = 3$

$\qquad 4d = 0$

$\qquad d = 0$

$6c + 0 = 3$

$c = \dfrac{1}{2}$

$\begin{pmatrix} a & b \\ c & d \end{pmatrix} = \begin{pmatrix} 0 & \dfrac{1}{2} \\ \dfrac{1}{2} & 0 \end{pmatrix}$

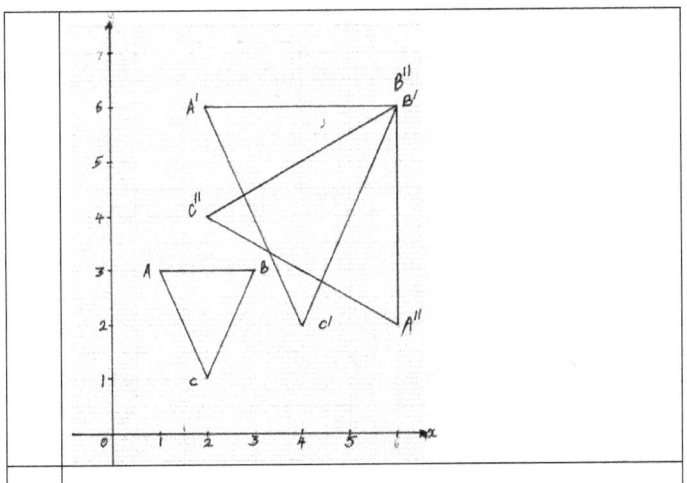

24.

$$\bar{x} = A + C\frac{\sum ft}{\sum f}$$

$$= 34.5 + \frac{10 \times (-193)}{180}$$

$$= 34.5 - 10.72$$

$$= 23.78$$

$$median = \frac{\left\{24.5 + \left(\frac{5}{30} \times 10\right)\right\} + \left\{24.5 + \left(\frac{6}{30} \times 10\right)\right\}}{2}$$

$$= \frac{26.17 + 26.50}{2} = \frac{52.67}{2}$$

$$= 26.335$$

$$(c) \, S.D = \sqrt{C^2 \left(\frac{\sum ft^2}{\sum ft}\right) - \left(\frac{\sum ft}{\sum ft}\right)^2}$$

$$= \sqrt{10^2 \left\{\frac{739}{180} - \left(\frac{-193}{180}\right)\right\}^2}$$

$$= \sqrt{100(4.106 - 1.150)}$$

$$= \sqrt{259.6}$$

$$= 17.19$$

CHAPTER SIX

1. Use logarithm tables only to evaluate $\sqrt[3]{\dfrac{\sin 6^0 38^1}{4.87 \times 0.03723}}$

2. Make N the subject of the formula

$$t = \frac{5P - N}{3N - P}$$

3. Solve the simultaneous equations below.

$$4x + y - 5 = 0 \ and \ x + 6y + 16 = 0$$

4. Expand $(2+2y)^5$. Hence find the value of $(2.02)^5$, correct to 4 decimal places when substitution for   y is upto y^4.

5. Without using calculators or mathematical tables. Express in surd form and simplify.

$$\frac{\cos 30}{\sin 45 + \tan 30}$$

6. Solve the given equation for values of $0^0 \le x \le 180^0$. .

$4\cos^2 x - 3\cos x = 6$.

7. There are three athletics A, B and C in a 100m race. A is twice as likely to win as B, while B is   twice as likely to win as C. find the probability that:-
   (a) A does not win the 100m race.
   (b) Either B or C wins the 100m race.

8.   The cost of printing a magazine is partly constant and partly varies as the number of pages of the   magazine. A magazine has 100 pages and the cost is ksh. 250 and if it has 50 pages the cost is ksh   150. Find the cost of printing a magazine with 300 pages.

9.   The price of a Nissan matatu at the end of 2003 was ksh 840,000/=. If it depreciates in value by 14% and 13% in the first and second years respectively and then by 10% in the subsequent years. Calculate its value at the end of 2010 to 4 s.f.

10.   Taps A and B can fill a tank in 4 and 9 hours respectively. Both taps are turned on for 2 hours   after which tap A is closed. Find how long tap B takes to fill the remaining part of the tank.

11.   Find the equation of a tangent to the curve y = 2x²-5x+2 at x = 3

12.   Find the centre and radius of a circle whose equation is $2x^2 + 2y^2 - 8x + 8y = 8$.

13.   Solve for x in 2 log x + log 6 = 2 + log 9

14.   Indian type A of rice costs ksh. 70 per kilogram while Egyptian type B of rice costs ksh 84 per kilogram. A shopkeeper mixes the two types of rice in the ratio 4:3 respectively. At what price   must he sell the mixture to make a profit of 26% per kilogram?

15.   Determine the amplitude, the period and phase angle of the graph $y = -4\sin\left(\frac{x}{4} - 55\right)$

16.   Calculate the interquatile range of the following numbers. 17, 34, 46, 58, 29, 78, 81, 85, 65, 77

17.   The nth term of the sequence is given by (3n+5).

(i)   Write down the first six terms of the sequence.

(ii)   Find the sum of the first 18 term of the series.

(iii)   Show that the sum of n terms is given by $S_n = \frac{1}{2}\left(13n + 3n^2\right)$

(iv) Determine the least value of n for which $S_n > 445$

18. Mr. John is a civil servant. He earns a basic monthly salary of ksh. 24 345, a house allowance of Ksh12,000 and medical allowance of ksh 2790, he is entitled to a personal relief of Ksh 1162 p.m, he has also an insurance scheme for which he pays premiums of Ksh 2300, he is entitled to a relief on the premiums at 15% of the premium paid. He is also a member of a cooperative society where he pays ksh 3000 towards his shares p.m.

Use the taxation table below to calculate his net salary per month, take to whole number.

Income (K£P.a)	Rate.(%)
1 -5808	10
5809 – 11280	15
11281 – 16752	20
16753 – 22224	25
Above 22224	30

19.   Four schools A, B, C and D are such that B is 94Km due North of A and C is an a bearing of 295 from A at a distance of 60km D is on a bearing of 310^0 from C and a distance of 42km.using a scale of 2cm to 20km, make an accurate scale drawing to show the relative positions of the   schools.

Find:

(i)   The distance and the bearing of B from C

(ii)   The distance and the bearing of D from B.

(iii)   The bearing of A from D

20. (a) Using a trapezoidal rule of 8 ordinates, estimate the area bounded by the curve
$y = 3x^2 + 4$ and the line x = 1, x = 5 and the x – axis.

(b) Find the exact area by integration in (a) above

(c) Determine the percentage error in area estimation in (a) above.

21. Construct triangle FGH with FG = 7cm, FH = 6.5 cm and GH = 5.2cm

(i) Construct the circum circle of triangle FGH and measure its radius.

(ii) In the same triangle FGH, the point Y moves such that it is at least 4.7cm from F and closer to GH than to GF. It lies within the confines of the triangle FGH. Indicate clearly the region in which Y must lie.

22. The figure below shows a plan of a roof with a rectangular base ABCD. AB = 20cm and BC = 12cm. the ridge PQ = 8cm and is centrally placed. The faces ADP and BCQ are equilateral   triangles. N is the midpoint of BC.

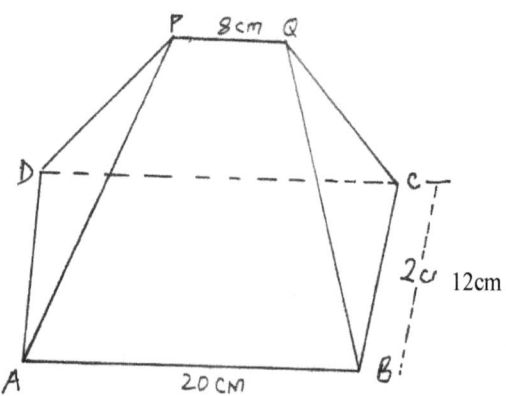

Calculate:

(a) The length of QN

(b) The altitude of P above the base.

(c) The angle between the planes ABQP and ABCD.

(d) The obtuse angle between the lines PQ and DB

23. Mokaya has 20 acres of land on which to grow maize and beans. For maize he has to employ one worker per acre and for beans he employs two workers per acre. The number of workers must not exceed 30. The total cost of growing beans is ksh 600 per acre and ksh 1000 per acre for maize. He cannot spend more than ksh 15000 altogether. He approximates the profit of maize to be ksh 400 per acre and ksh 600 per acre of beans.

   (a) Form all the inequalities to represent the information above. Take x to represent acres for maize and y beans.

   (b) On the grid provided, draw the inequalities to show the wanted region.

   (c) Use your graph in (b) above to determine the number of acre of maize and beans he has to plant in order to maximize the profit and find the profit.

24. The position of two cities P and Q are (39^0S, 41^0E) and (39^0S, 129^0W) respectively.

   (a) Find the difference in longitude between P and Q.

   (b) Given that the radius of the earth is 6370km, calculate the distance between P and Q

   (i) In kilometers (take $\pi = \dfrac{22}{7}$)

   (ii) In nautical miles

   (c) Another city R is 870km east of city Q and on the same latitude as cities P and Q. Find the   longitude of city C.

93

SOLUTIONS TO CHAPTER SIX

No.	Working	
1.	No $\sin 6^0 38^1$ 4.87 0.03723 8.604×10^{-1}	log : $\overline{1}.0626$ 0.6875 + $\overline{2}.5709$ $\overline{1}.2584 \rightarrow \overline{1}.2584$ $\overline{1}.8042 \times \frac{1}{3}$ $\overline{3}+2.8042$ 3 $\overline{1}.9347$	-
	$= 0.8604\text{ans}$		
2.	$t = \dfrac{5P - N}{3N - P}$ $t(3N - P) = 5P - N$ $3tN - tP = 5P - N$ $3tN + N = 5P + tP$ $N(3t + 1) = 5P + tP.$ $N = \dfrac{5P + tP}{3t + 1}$		
3.	$4x + y = 5$ $x + 6y = -16$ $24x + 6y = 30$ $x + 6y = -16$ $23x = 46$ $x = \dfrac{46}{23} = 2$ $4x + y = 5$ $8 + y = 5$ $y = -3$		
4.	$(2+2y)^5=2^5+5(2)^4 2y)+10(2)^3(2y)^2+10(2)^2(2y)^3+5(2)(2y)^4+(2y)^5$ $[=32+160y+320y^2+320y^3+160y^4+32y^5](2.02)^5=(2+0.02)^5y=0.01$		

94

	$=32+160(0.01)+320(0.01)^2+320(0.01)^3+160(0.01)^4+32(0.10)^5$ $=32+1.6+0.032+0.00032+0.0000016$ $=33.6323216$ $\cong 33.6323$
5.	$\dfrac{\cos 30}{\sin 45 + \tan 30}$ $\dfrac{\dfrac{\sqrt{3}}{2}}{\dfrac{1}{\sqrt{2}}+\dfrac{1}{\sqrt{3}}}$ $=\dfrac{\dfrac{\sqrt{3}}{2}}{\left(\dfrac{1}{\sqrt{2}}+\dfrac{1}{\sqrt{3}}\right)\left(\dfrac{\sqrt{2}-\sqrt{3}}{\sqrt{2}-\sqrt{3}}\right)}$ $=\dfrac{\dfrac{\sqrt{3}}{2}}{\dfrac{\sqrt{2}-\sqrt{3}+\sqrt{2}-\sqrt{3}}{\left(\sqrt{2}\right)-\left(\sqrt{3}\right)}}$ $=\dfrac{\dfrac{\sqrt{3}}{2}}{\dfrac{2\sqrt{2}-2\sqrt{3}}{-1}}=\dfrac{\dfrac{\sqrt{3}}{2}}{2\sqrt{3}-2\sqrt{2}}$ $=\dfrac{\dfrac{\sqrt{3}}{2}}{\dfrac{2\sqrt{2}-2\sqrt{3}}{1}}=\dfrac{\dfrac{\sqrt{3}}{2}}{2\sqrt{3}-2\sqrt{2}}$ $=\dfrac{\dfrac{\sqrt{3}}{2}\left(2\sqrt{3}+2\sqrt{2}\right)}{\left(2\sqrt{3}-2\sqrt{2}\right)\left(2\sqrt{3}+2\sqrt{2}\right)}$ $=\dfrac{\dfrac{\sqrt{3}}{2}\left(2\sqrt{3}+2\sqrt{2}\right)}{\left(2\sqrt{3}\right)^2-\left(2\sqrt{2}\right)^2}$ $=\dfrac{3+\sqrt{6}}{4}$

6.	$4\cos^2 x - 3\cos x = 6$
	$4\cos x^2 - 3\cos x - 6 = 0$
	$x = \dfrac{-b \pm \sqrt{b^2 - 4ac}}{2a}$
	$= \dfrac{3 \pm \sqrt{(-3)^2 - (4 \times 4 \times -6)}}{2 \times 4}$
	$= \dfrac{3 \pm \sqrt{9 + 96}}{8} = \dfrac{3 \pm \sqrt{105}}{8}$
	$= \dfrac{3 \pm 10.25}{8} = \dfrac{13.25}{8} \, or \, \dfrac{-7.28}{8}$
	$\therefore x = -0.90625 \, or \, 1.65625 \; ignore$
	$\therefore x = \cos^{-1}(0.90625)$
	$= 154.9921668^0$
	$\cong 155^0$
7.	$A = 2x, B = x, C = \dfrac{1}{2}x$
	$\therefore A : B : C = 2x : x : \dfrac{1}{2}x$
	$= 4x : 2x : x$
	$(a) \, P(A \, not \, winning) = P(A) = 1 - \dfrac{4}{7} = \dfrac{3}{7}$
	$(b) \, P(B \, or \, C \, winning) = P(B) or P(C)$
	$= \dfrac{2}{7} + \dfrac{1}{7} = \dfrac{3}{7}$
8.	Equation C = a + bn
	$250 = a + 100b$
	$\dfrac{150 = a + 50\,b}{100 = \quad 50\,b}$

	$\therefore b = \dfrac{100}{50} = 2$
	$150 = a + 30(2)$
	$= a + 100$
	$a = 50, b = 2$
	$c = 50 + 2n$
	$= 50 + 2(300)$
	$= 50 + 600$
	$= 650$
	Cost of magazine with 300pages is ksh 650
9.	At the end of 2004, value of Nissan
	$= \dfrac{86}{100} \times 840,000$
	$= ksh722,400$
	At the end of 2005 value of Nissan
	$= \dfrac{87}{100} \times 722400$
	$= ksh628488$
	At the end of 2010
	$A = 628488\left(1 - \dfrac{10}{100}\right)^5$
	$= 628488(0.9)^5$
	$= 628488 \times 0.59049$
	$= 371115.8791$
	$\cong ksh371,100$
10.	In two hours boths fills

	$\left(\dfrac{1}{4}\times 2+\dfrac{1}{9}\times 2\right)=\dfrac{1}{2}+\dfrac{2}{2}=\dfrac{13}{18}$ *filled* *fraction not filled* $1-\dfrac{13}{18}$ $\qquad\qquad\qquad =\dfrac{5}{18}$ B *alone fills* $\dfrac{1}{9}$ *in* 1 *hour to fill* $\dfrac{5}{18}$ $=\dfrac{5}{18}\times\dfrac{9}{1}$ $=10\ \ hours$
11.	$y = 2x^2 - 5x + 2$ $Gradient = \dfrac{dy}{dx}= 4x-5$ $\qquad\qquad = (4 \times 3) - 5 = 7$ $\therefore$ Equation of tangent $y = mx + c$ at $x = 3$, $y = (2x2^2) - (5x3) + 2$ $\qquad\qquad\qquad\qquad = 8 - 15 + 2 = -5$ $\therefore y = mx + c$ at $(3, -5)$ gr $= 7$ $-5 = (7 \times 3) + c$ $c = -5 - 21 = -26$ $y = 7x - 26$
12.	$2x^2+2y^2-8x+8y=8$ $X^2+y^2+4y-4x=4$ $X^2-4x+y^2+4y=4$ $X^2-4x+4+y^2+4y+4=4+4+4$ $(x-2)^2+(y+2)^2=12$ $\therefore$ centre $(2, -2)$ radius $= \sqrt{12}$ $\qquad\qquad\qquad = 3.464$
13.	$2\log x + \log 6 = \log 100 + \log 9$ $\log x^2 + \log 6 = \log 100 + \log 9$ $\log x^2 = \log 100 + \log 9 - \log 6$ $\log x^2 = \log\left(\dfrac{100\times 9}{6}\right)$ $x^2 = \dfrac{100\times 9}{6}$ $= 150$ $= \pm\sqrt{150}$ $= 12.25$

14.	Cost of mixture per kg $$\frac{(4 \times 70) + (84 \times 3)}{4 + 3}$$ $$= \frac{280 + 252}{7}$$ $$= \frac{532}{7}$$ $$= ksh.76$$ $$profit\ of\ 26\% = \frac{126 \times 76}{100}$$ $$= ksh95.76$$
15.	$$y = -4\sin\left(\frac{x}{4} - 55\right)$$ $$\therefore amplitude = 4$$ $$period = \frac{360}{\frac{1}{4}} = 1440^0$$ $$phase\ angle = 220^0$$
16.	17,29,34,46,58,65,77,78,81,85 $Q_1 = 34$ (lower quatile) $Q_3 = 78$ (upper quatile) Interquatile range $Q_3 - Q_1$ $= 78 - 34$ $= 44$
17.	(i) $(3n+5)$, n=1, 2, 3, 4, 5, 6 Tarry 8, 11, 14, 17, 20, 23. (ii) $$Sn = \frac{n}{2}(2a + (n-1)d)$$ $$S_{18} = \frac{18}{2}(2 \times 8 + (18-1)3)$$ $$= 9(16 + 17 \times 3)$$ $$= 9(16 + 51)$$ $$= 9(67)$$ $$= 603$$ (iii)

$$Sn = \frac{n}{2}\left(2a + (n-1)d\right)$$

$$= \frac{n}{2}\left(2 \times 8 + (n-1)3\right)$$

$$= \frac{n}{2}\left(16 + 3n - 3\right)$$

$$\frac{n}{2}\left(13 + 3n\right)$$

$$= \frac{1}{2}\left(13n + 3n^2\right)$$

(iv)

$$\therefore \frac{1}{2}\left(13n + 3n^2\right) > 445$$

$$13n + 3n^2 > 890$$

$$3n^2 + 13n - 890 > 0$$

$$n = \frac{-b \pm \sqrt{b^2 - 4ac}}{2a}$$

$$n = \frac{-13 \pm \sqrt{(13)^2 - (4 \times 3 \times -890)}}{2 \times 3}$$

$$= \frac{-13 \pm \sqrt{169 + 10680}}{6}$$

$$= \frac{-13 \pm \sqrt{10849}}{6}$$

$$= \frac{-13 \pm 104.1585}{6}$$

$$= \frac{-117.1585}{6} \, or \, \frac{91.1585}{6}$$

$$= -19.526 \, or \, 15.193$$

$$\therefore n = 15$$

18.	Taxable income

Taxable income
$= (24345 + 1200 + 2790)$
$= ksh.39135\, p.m$

Total K£$= 39135 \times \frac{12}{20}$

$=$k£ 23481
Taxation
1st $5808 \times \frac{10}{100} = 580.8$

2nd $5472 \times \dfrac{15}{100} = 820.8$

3rd $5472 \times \dfrac{20}{100} = 1094.4$

4th $5472 \times \dfrac{25}{100} = 1368.0$

5th balance $= 23481 - 22,224$
$\qquad\qquad = 1257$

$\Rightarrow 1257 \times \dfrac{30}{100} = 377.1$

Total tax due
$= k£(580.8 + 820.7 + 1094.4 + 1368..1 + 377.1)$
$= K£\ 4241.1$

Insurance relief
$= \dfrac{15}{100} \times 2300 \times \dfrac{12}{20} = k£\ 207$

Personal relief
$= 1162 \times \dfrac{12}{20} = K£464.8$

Total relief $= k£671.8$

Tax payable per year
$= 4241.1 - 671.8$
$= k£\ 3569.3$

His total deductions p.a are

$k£\ \left(3569.3 + \dfrac{2300 \times 12}{20} + \dfrac{3000 \times 12}{20}\right)$

$= (3569.3 + 1380 + 1800)$
$= 6749.3 k£$

His deductions per month

$= \left(\dfrac{6749.3 \times 20}{12}\right)$

$= 11248.83$

His net salary per month
$= 39135.00$
$\quad\underline{11248.83}$
$\underline{Ksh\ 27886.}$

| 19. | |

(i) 88 km $\pm$ 1 and $049^0 \pm 1$

(ii) 96km $\pm$ and $254^0 \pm 1$

(iii) 90 + 31

= 121 ± 2^0

---

20. (a) $y = 3x^2 + 4$

x	1	1.5	2	2.5	3	3.5	4	4.5	5
y	7	10.75	16	22.75	31	40.75	52	64.75	79

$$area = \frac{h}{2}\{(y_0 + y_n) + 2(y_1 + y_2 + \ldots\ldots y_{n-1})\}$$

$$= \frac{0.5}{2}\{(7 + 79) + 2(10.75 + 16 + 22.75 + 31 + 40.75 + 52 + 64.75)\}$$

$$= \frac{0.5}{2}\{(86) + 2(238)\}$$

$$= \frac{0.5}{2}\{86 + 476\}$$

$$= 140.5$$

$$\int_1^5 (3x^2 + 4)$$

$$= [x^3 + 4 \times 5]$$

$$= [5^3 + 4 \times 5] - [1^3 + 4(1)]$$

$$= [125 + 20] - [1 + 4]$$

$$= 145 - 5$$

$$= 140$$

(c)

$Abs.error = /140 - 140.5/ = 0.5$

$$\%Error = \frac{0.5}{140} \times 100$$

$$= 0.3571\%$$

---

21. (a)

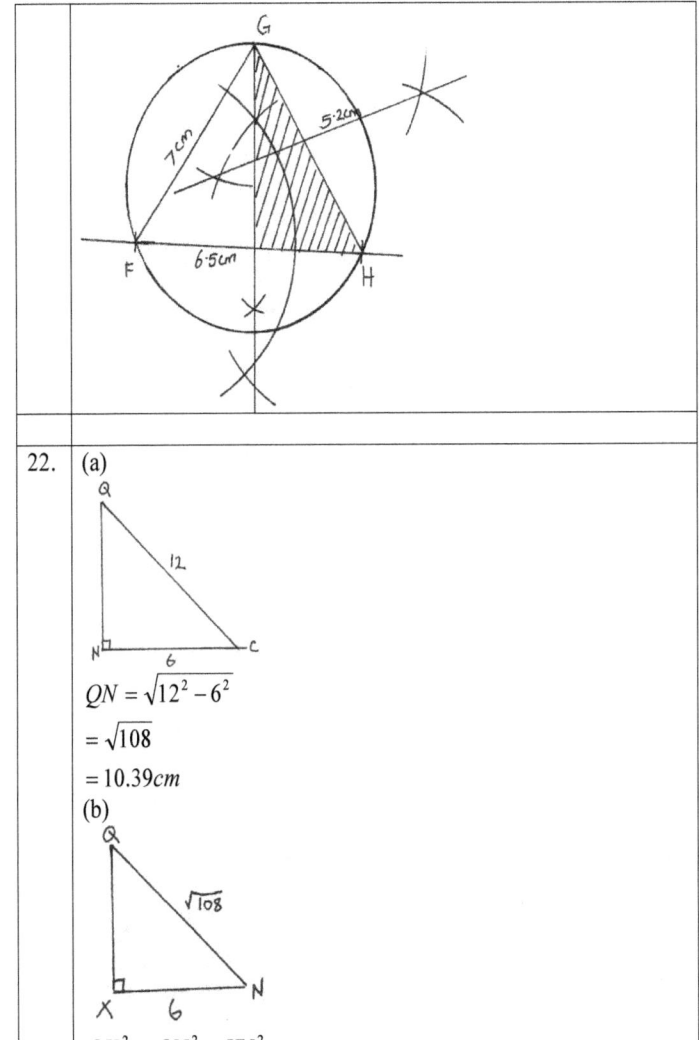

22. (a)

$QN = \sqrt{12^2 - 6^2}$

$= \sqrt{108}$

$= 10.39cm$

(b)

$QX^2 = QN^2 - XN^2$

$= \sqrt{108}^2 - 6^2$

$= 108 - 36$

$QX = \sqrt{72}$

$= 8.485$

(c)

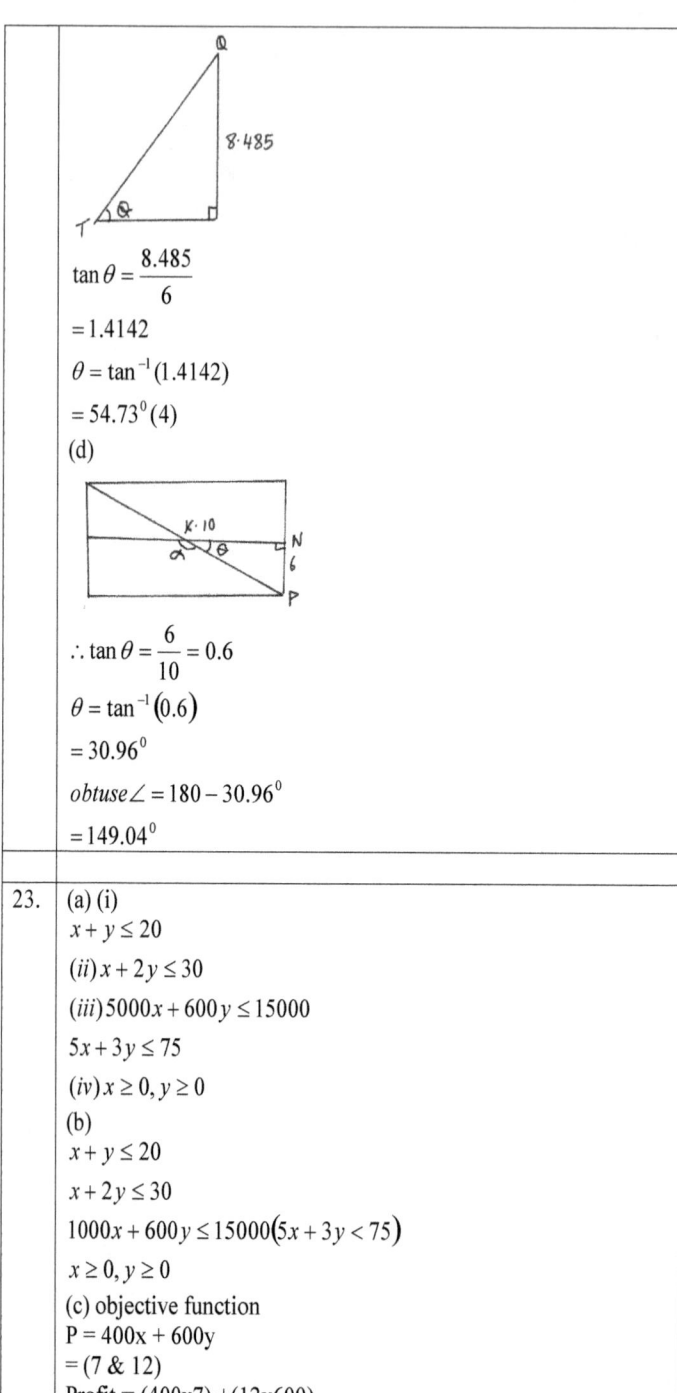

$$\tan \theta = \frac{8.485}{6}$$

$$= 1.4142$$

$$\theta = \tan^{-1}(1.4142)$$

$$= 54.73^0 (4)$$

(d)

$$\therefore \tan \theta = \frac{6}{10} = 0.6$$

$$\theta = \tan^{-1}(0.6)$$

$$= 30.96^0$$

$$obtuse \angle = 180 - 30.96^0$$

$$= 149.04^0$$

23.	(a) (i) $x + y \leq 20$ $(ii) x + 2y \leq 30$ $(iii) 5000x + 600y \leq 15000$ $5x + 3y \leq 75$ $(iv) x \geq 0, y \geq 0$ (b) $x + y \leq 20$ $x + 2y \leq 30$ $1000x + 600y \leq 15000(5x + 3y < 75)$ $x \geq 0, y \geq 0$ (c) objective function P = 400x + 600y = (7 & 12) Profit = (400x7) +(12x600) =2800 + 7200

	$= 10{,}000/=$
24.	(a) Longitude (angle) difference $= 41 + 1290$ $= 170^0$ (b) (i) Distance in km $= \dfrac{170}{360} \times 2 \times \dfrac{22}{7} \times 6370 \cos 39^0$ $= \dfrac{17}{80} \times 44 \times 910 \times 0.7771$ $= 14693.23 km$ (ii) $= 60 \times 170 \cos 39$ $= 60 \times 170 \times 0.7771$ $= 7926.42 NM.$ (c) $Dis\tan ce = \dfrac{\theta}{360} \times 2\pi$ $\therefore 870 = \dfrac{\theta}{360} \times 2 \times \dfrac{22}{7} \cos 39^0 \times 6370$ $\theta = \dfrac{870 \times 360 \times 7}{2 \times 22 \times 6730 \cos 39^0}$ $= \dfrac{870 \times 360 \times 7}{44 \times 6370 \times 0.7771}$ $= \dfrac{31320}{3111.5084}$ $Q = 10.07^0$ $city\,C's\,longitude = 129 - 10.07^0$ $\qquad\qquad\qquad = 118.93^0$ $C = (39^0 S, 118.93^0)$

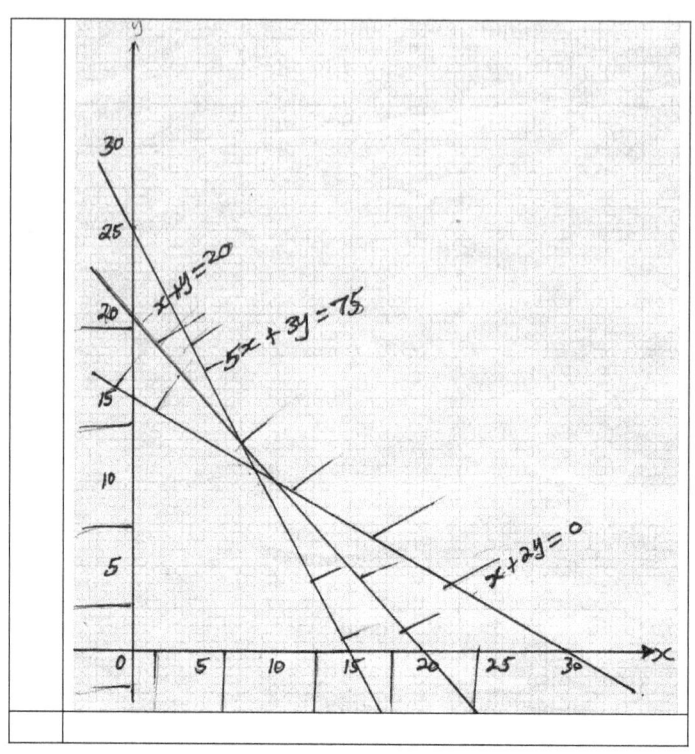

CHAPTER SEVEN

1. a) Write the odd numbers in descending order between 1 and 10 inclusive.

    b) Round off the number formed to 3 significant figures.

    c) Find the total value of digit 7 in the new number in 2(b) above.

2. The L.C.M. of three numbers is 360 and the G.C.D of the same numbers is 2. If one of the numbers is 40. find the other 2 numbers.

3. Mwangi and Otieno live 60km apart. Mwangi leaves his home at 7.00am cycling towards Otieno's house at 20km/h. Otieno leaves his home at 8.00am cycling towards Mwangi's house at 8mk/h

    a) At what time did they meet?

    b) How far is the meeting point from Mwangi's house?

4. Divide $10^{10}/_{27}$ by $2^7/_9$ then add the result to the product of $6^2/_3$ and $^4/_{25}$. Find three quarters of the result and leave your answer as a fraction.

5. Express the following recurring decimal as a fraction in its simplest form.

    $0.1\dot{5}\dot{3}$

6. Use the square, cube root, and reciprocal table to evaluate to 4 d.p

    $$\frac{\sqrt[3]{0.008}}{0.375} - \frac{10}{37.5^2}$$

7. A two digit number is such that the sum of its digits is 13. The product of twice the tens digit and the units digit minus the original number is 22. Find the number.

8. Without using tables or calculators evaluate. Give your answer as a mixed number.

    $$\frac{(-8)\times 4 + 156 \div 4 \; of \; (-36 + 30)}{(-5) - (-8)\times 2 + 6}$$

9. Solve for x in the following equation.

$$\left(\frac{1}{4}\right)^{x-2} = 2^{x+2}$$

10. Three points on a map A, B, and C are such that the bearing of B and C from A are $\theta\,60^0$ and 150^0 respectively. If the distance AB is 0.45m and BC is 0.75m, determine the distance AC.

11. 0.05 litres of water is poured into an empty measuring cylinder. A piece of metal with mass 135g is put into the cylinder. If the density of the metal is $9600kg/m^3$, find the new reading of the cylinder.

12. A newspaper vendor has 30 bank notes with a total value of Sh. 900, if the notes are either in Sh. 20 or Sh. 50 denominations, how many of each does he have?

13. The length of a rectangle was increased by 30% while its width decreased by 15%. Determine the % change in the area of the rectangle.

14. Find the integral values of x which satisfy the following inequalities.
$$2\,(2 - x) < 4x - 9 < x + 11.$$

15. The figure below shows a solid made by placing two equal regular tetrahedra.

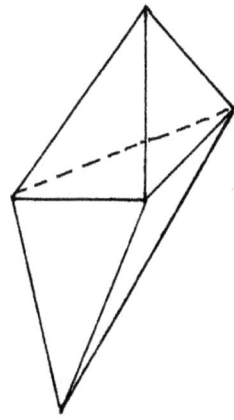

a) Draw the net of the solid.

b) If each face is an equilateral triangle of side 5cm. Find the surface area of the solid to 4 s.f

16. Using a ruler and a pair of compasses only, draw a parallelogram ABCD in which AB = 8cm, BC = 6cm and <BAD = 75°. Drop a perpendicular from D to meet AB at N. Determine the length DN.

17. Two business men Achaki and Mkazi contributed Ksh. 128, 000 and Ksh. 112,000 respectively to start a business. They agreed to share the profits as follows;

         30% shared equally

         30% shared in ratio of contributors

         40% retained for running business.

Their profit for the year 2008 was Ksh. 86, 400

Calculate.

a) The amount shared equally.

b) The total amount received by each partner.

c) The amount retained for running the business.

18.   Rates of tax in operation in 2009 are as given in the table below:-

K£ pa	Rate of tax %
1 – 5208	10
5209 – 9744	25
9745 – 14292	20
14293 – 18840	15
Over 18840	30

a)   Mr. Rono pays Sh. 5400 as P.A.Y.E monthly. He was entitled to house allowance of Kshs. 9000pm and getting a monthly tax relief of Sh. 1093. Calculate his monthly basic salary.

b)   Mr. Rono's other deduction per month were.

Co-operation society contribution   Sh. 2000.

Loan repayment   Sh. 2500

Calculate his Net salary per month

19.   A Rhombus has its vertices as PQRS. The co-ordinates of the vertex P and Q of the rhombus are P(-1, 3) and Q (2, 4). The diagonal QS and PR meet at point M. Given that the equation of the line PR is $y = x + 4$.

a)   Find the equation of the diagonal QS.

b)   Find the co-ordinates of the midpoint M of QS.

c)   Find the co-ordinates of the points R and S.

d)   Plot on the grid provided the rhombus PQRS.

20.   The table below shows marks scored by students in a maths test.

Marks	≤ 10	≤ 20	≤ 30	≤ 40	≤ 50	≤ 60	≤ 70	≤ 80	≤ 90	≤ 100
No. of students	2	5	10	18	27	33	38	41	43	44

a)   Prepare a frequency distribution table from the above table.

b)   i)   State the modal class.

      ii)   Calculate the mean and the median.

c)   Find the mark if the percentage pass was $59^{1}/_{11}\%$.

21.   During a school assembly one Friday, three scouts, Morgan, Boris and Zeddy each stood 6m away from the foot of a vertical flag post 6m high Their bearings from the flag post were 060^{0}, 110^{0} and 220^{0} respectively.

   a)   Draw a sketch of their relative positions.
   b)   Using a scale of 1cm: 1m make an accurate drawing of their positions.
   c)   Find the bearing of:
      i)   Morgan from Zeddy
      ii)   Boris from Morgan.
   d)   Find the distance of Zeddy from Boris.
   e)   Find the angle of elevation of the top of the flag post from:
      i)   Morgan.
      ii)   Boris
iii) Zeddy

22.

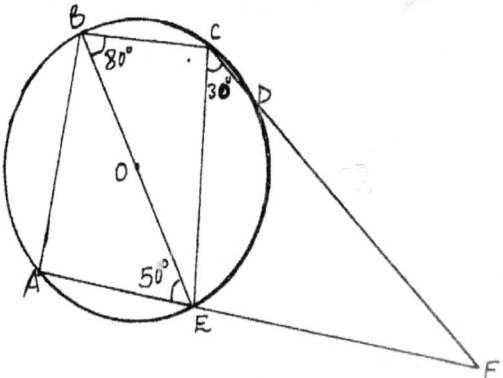

In the figure above, O is the centre of the circle. Angle AEB = 50^{0}, angle EBC = 80^{0} and angle ECD = 30^{0}. Giving reasons, calculate:

   i)   Angle CDE
   ii)   Angle DFE
   iii)   Obtuse Angle COE

iv) Angle ADE

23.   A tank with hemispherical dome-shaped base of radius 14cm was filled with water up to

the top (tangent) of the hemispherical dome shape base as shown below. ($\pi = \dfrac{22}{7}$)

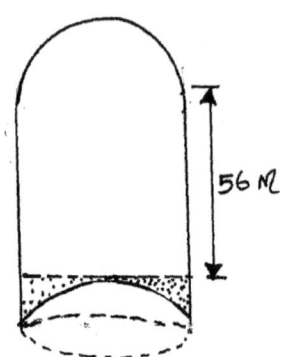

56 ᴿℓ

Calculate:

a)   The surface area of the tank in contact with the water.

b)   The volume of water in litres.

c)   A feeder pipe of diameter 14cm supplies water to this empty tank at the rate of $40cm^3$
/sec. Calculate the time taken for the tank to be filled completely in hours.

24.   In an n-sided polygon, two angles are Right angles and each of the remaining angles is
150^0.

    a)   Find the value of **n** hence the sum of interior angles of this polygon.

    b)   Name the polygon.

    c)   Find the area of a regular octagon of sides 4cm to 5 s.f.)

SOLUTIONS TO CHAPTER SEVEN

1	a) 9, 7, 5, 3, 1 b) 97,500 c) 7 x 1000 = 7,000
	Total
2	L.C.M = 360 = 2^3 x 3^2 x 5 G.C.D = 2 40 = 2^3 x 5 x_1 = 2 x 3^2 = 18 x_2 = 2 x 5 = 10
	Total

3	a) Relative speed = (20 + 8) = 28km/h
	    Dist Mwangi traveled by 8.00 = (20 x 1) = 20km
	    Remaining dist. (60 - 20) = 40km
	
	    Time taken to meet = $\dfrac{40}{28}$ = 1¾ hrs
	
	       i.e. 1hr ($^3/_7$ x 60) mins = 1hr 25mins
	    Time of meeting 8.00
	                  + 1.25
	                  9.25a.m
	 b) Dist = speed x time
	
	     = 20 x $\dfrac{40}{28}$
	
	     = $\dfrac{800}{28}$ km = 28.57km

	Total
4.	$$\left\{10\frac{10}{27} \div 2\,^7/_9 + \left(6\,^2/_3 \times \,^4/_{25}\right)\right\} \times \,^3/_4$$ $$= \left\{\frac{280}{27} \times \frac{9}{25} + \left(\frac{20}{3} \times \frac{4}{25}\right)\right\} \times \,^3/_4$$ $$= \left\{\frac{56}{15} + \frac{16}{15}\right\} \times \,^3/_4$$ $$= \frac{72}{15} \times \,^3/_4$$ $$= \,^{18}/_5 = 3\,^3/_5$$
	Total
5.	$r = 0.153153$ $100r = 153.153153$ $1000r - r = 153.153153$ - 0.153153

	$999r = 153$	
	$r = \dfrac{153}{999}$	
	$r = \dfrac{17}{111}$	
		Total
6	$\dfrac{\sqrt[3]{0.008}}{0.375} = 0.2 \times \dfrac{1}{0.375}$	
	$\quad = 0.2 \times 2.667 = 0.5334$	
	$\dfrac{10}{37.5^2} = 10 \times \dfrac{1}{37.5^2}$	
	$\quad = 10 \times \dfrac{1}{1406}$	
	$\quad = 10 \times 0.0007172 = 0.007172$	
	$Difference \, (0.5334 - 0.007172)$	
	$\quad = 0.526268$	
	$\quad = 0.5262$	
		Total
7	Let number be $(10x + y)$	
	$x + y = 13$ hence $x = 13 - y$.......... (i)	
	also $2xy - (10x + y) = 22$	
	$\quad 2xy - 10x - y = 22$.............. (ii)	
	Hence,	
	$2(13 - y) y - 10 (3 - y) - y = 22$	
	$26 - 2y^2 - 130 - 10y - y = 22$	
	$26y + 9y - 2y^2 \, 130 - 22 = 0$	
	$-2y^2 + 35y - 152 = 0$	
	$y = \dfrac{-35 \pm \sqrt{35^2 - 4(-2)(-152)}}{2(-2)}$	
	$y = \dfrac{-35 \pm \sqrt{1225 - 1216}}{-4}$	
	$y = -\dfrac{35 \pm \sqrt{9}}{-4} = \dfrac{-35 \pm 3}{-4}$	
	$y = \dfrac{-32}{-4} = 8, \; y = \dfrac{-38}{-4} = -9.5$	
	$Hence \; y = 8 \; \& \; x = 5, \; Number \; is \, 58$	
		Total

8	$\dfrac{(-8)\times 4 + 156 \div 4 \; of \; (-6)}{(-5) - (-8)\times 2 + 6}$
	$= \dfrac{-32 + 156 \div (-24)}{(-5) - (-8)\times 2 + 6}$
	$= \dfrac{-32 - 65}{-5 + 16 + 6}$
	$= \dfrac{-97}{17}$
	$= -5\dfrac{12}{17}$
	Total
9	$(2^{-2})^{x-2} = 2^{x+2}$
	-2 (x - 2) = x + 2
	-2x + 4 = x + 2
	-3x = -2
	x = $^2/_3$
	Total
10	
	<A = 90^0
	Therefore, 0.45^2 + AC^2 = 0.75^2
	Therefore, AC2 = 0.36
	AC = 0.6m
	Total
11	Volume of water in cylinder
	= 0.05 litres = 50cm3
	Volume of metal = $\dfrac{0.135kg}{9600kg/m^3}$ Or $\dfrac{135g}{9.6g/cm^3}$
	= 14.06cm^3
	New volume = (50 + 14.06)cm^3
	= 64.06cm^3
	Total
12	Let Ksh 20 notes be x
	and Ksh 50 notes be y
	number of notes x + y = 30
	value of notes 20x + 50y = 900

	solving $20x + 50y = 900$
	                 $20x + 20y = 600$
	                      $30y = 300$
	                         $y = 10$
	then $x + 10 = 30$, $x = 20$
	vendor has 20, Ksh 20 notes and 30, Ksh 50notes

	Total

13.	Initial area $= L \times W = LW$
	New area $= 1.3l \times 0.85w$
	            $= 1.105lw$
	Area Diff $= (1.105 - 1) = 0.105$
	% Increase $= \dfrac{0.105}{1} \times 100$
	        $= 10.5\%$

	Total
14.	$4 - 2x < 4x - 9$ $4x - 9 < x + 11$ $4 + 9 < 4x + 2x$ $5x < 20$ $13 < 6x$ $x < 4$ $1\tfrac{3}{6} < x < 4$ or $2\tfrac{1}{6} < x < 4$ Ans 3
	Total

15	$Surface\,area = 6\left(\sqrt{s(s-a)(s-b)(s-c)}\right)$
	
	$S = \tfrac{1}{2}(5+5+5) = 7.5$
	
	$S.A = 6\left(7.5(7.5-5)(7.5-5)(7.5-5)\right)^{\!1/2}$
	
	$S.A = 6\sqrt{(7.5 \times 2.5^3)}$
	
	$S.A = 6 \times \sqrt{(7.5 \times 15.625)}$
	
	$S.A = 6 \times \sqrt{(117.1875)}$
	
	     $= 6 \times 10.825$
	
	     $\cong 64.95cm^3$

	Total

16	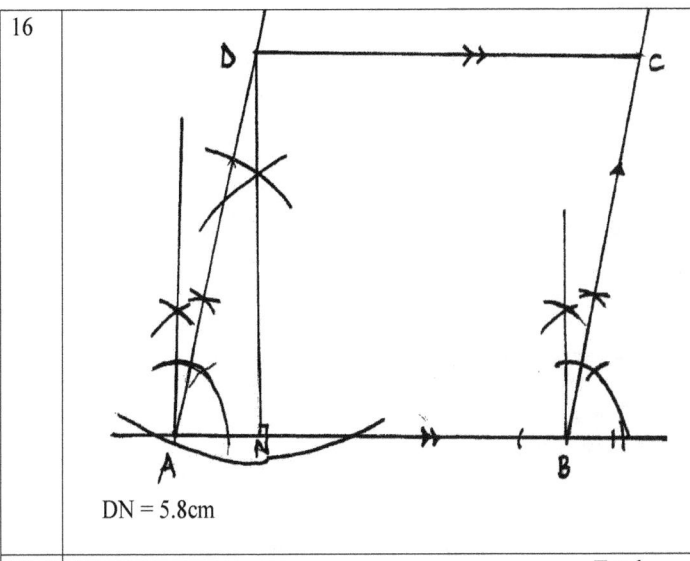 DN = 5.8cm
	Total

17	a) $\dfrac{30}{100}$ x 86,400 = 25,920 Each received ½ x 25, 920 = Kshs 12,960 b) Ratio 128000 : 112000 = 8 : 7 (A :M) $\dfrac{8}{15}$ x 25, 920 = 13,824 $\dfrac{7}{15}$ x 25, 920 = 12, 096 Achoki (13, 824 + 12, 960) = 26, 784Ksh Mkazi (12,096 + 12,960) = 25,056Ksh c) $\dfrac{40}{100}$ x 86, 400 = Ksh. 34,560
	Total

18	a) 5208 x $\dfrac{10}{100}$ x 20 = 10416 4536 x $\dfrac{15}{100}$ x 20 = 13608 4548 x $\dfrac{20}{100}$ x 20 = 18192 4548 x $\dfrac{25}{100}$ x 20 = 22740

$$X \times \frac{30}{100} \times 20 = 12960$$

$$x = \frac{12960}{6} = 2160$$

Total tax = 77916
           +13116
           Ksh. 64800

Total income (5208 + 4536 + 4548 + 4548 + 2160)
                K£ 21,000 p.a.

Monthly basic salary = $\frac{21,000 \times 20}{12}$ - 9000

$$= 35,000 - 9000$$
$$= \text{Ksh. } 26,000$$

b) Total deductions (5400 + 2000 + 2500)
                     = Ksh. 9900

Net salary (p.m) = 35,000 – 9900
                  = Ksh. 25,100

		Total
19	a) Gradient of QS = -1 $\frac{y-4}{x-2} = -1$ y – 4 = -1 (x - 2) y = -x + 6 b) Co-ordinates of M y = -x + 6 y = x + 4 solving –x + 6 = x + 4 -2x = -2 x = 1 y = 1 + 4 y = 5 M (1, 5) c) Mid point of QS is M (1, 5) x - co-ordinate $\frac{x_1 + x_2}{2} = 1$ $2 + x_2 = 2$ $x_2 = 0$ y- co-ordinate $\frac{y_1 + y_2}{2} = 5$ $4 + y_2 = 10$ $y_2 = 6$ Therefore, S (0, 6) Mid point of PR is M (1,5) x – co=ordinate $\frac{x_1 + x_2}{2} = 1$ $-1 + x_2 = 2$ x2 = 3	

y – co-ordinate $\frac{y_1 + y_2}{2} = 5$

$3 + y_2 = 10 \qquad y_2 = 7$

R (3, 7)

d)

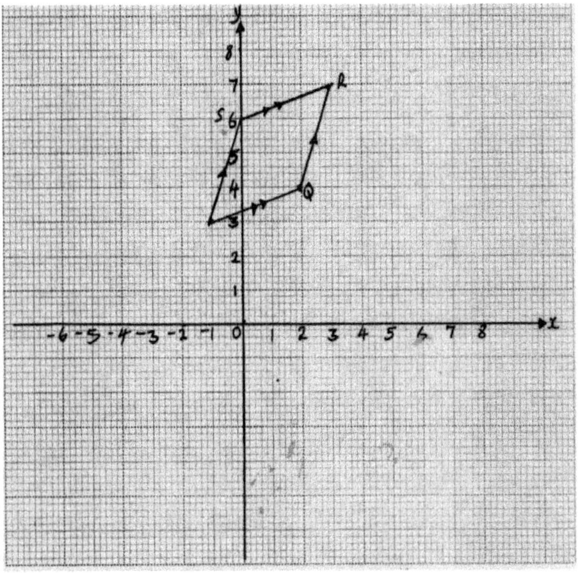

Total

20

Class	Mid pt x	f	fx	cf
0-10	5.5	2	110	2
11-20	15.5	3	46.50	5
21-30	25.5	5	127.50	10
31-40	35.5	8	284.0	18
41-50	45.5	9	409.5	27
51-60	55.5	6	333.0	33
61-70	65.5	5	327.5	38
71-80	75.5	3	226.5	41
81-90	85.5	2	171.0	43
91-100	95.5	1	95.5	44
		44	2032	

b) i) Modal class 41 - 50

ii) Mean = $\frac{2032}{44}$ = 46.1818

Median = $40.5 + \left(\frac{22.5 - 18}{9}\right)10$

= 45.5

	c) $59\frac{1}{11}\%$ of $44 = 26$ students. Therefore failed are $18 =>$ pass mark $= 40\%$
	Total

21	a)
	
	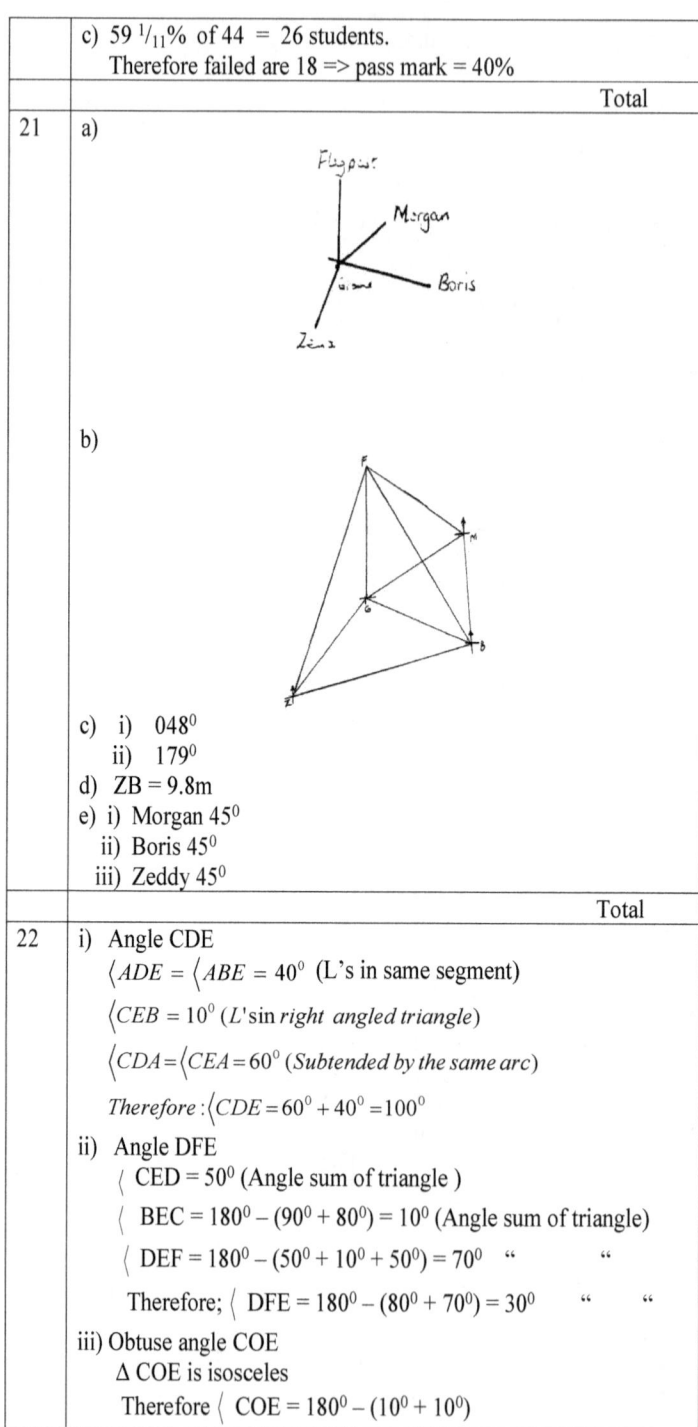
	
	b)
	
	c) i) 048^0
	   ii) 179^0
	d) $ZB = 9.8$m
	e) i) Morgan 45^0
	   ii) Boris 45^0
	   iii) Zeddy 45^0

	Total
22	i) Angle CDE $\langle ADE = \langle ABE = 40^0$ (L's in same segment) $\langle CEB = 10^0$ (L'sin *right angled triangle*) $\langle CDA = \langle CEA = 60^0$ (*Subtended by the same arc*) *Therefore* : $\langle CDE = 60^0 + 40^0 = 100^0$ ii) Angle DFE $\langle$ CED $= 50^0$ (Angle sum of triangle) $\langle$ BEC $= 180^0 - (90^0 + 80^0) = 10^0$ (Angle sum of triangle) $\langle$ DEF $= 180^0 - (50^0 + 10^0 + 50^0) = 70^0$ " " Therefore; $\langle$ DFE $= 180^0 - (80^0 + 70^0) = 30^0$ " " iii) Obtuse angle COE Δ COE is isosceles Therefore $\langle$ COE $= 180^0 - (10^0 + 10^0)$

	$= 160^0$ (Angle sum of triangle) iv) Angle AOE $\quad \langle$ ABE $= 180^0 - (50^0 + 90^0) = 40^0$ (Angle sum of triangle) $\quad \langle$ ADE $= 40^0 = \langle$ ABE (Subtended by same arc) Therefore ADE $= 40^0$
	Total
23	a) Surface area in contact with water $= \frac{1}{2} \times 4\pi r^2 + 2\pi rh$ $\qquad = (\frac{1}{2} \times 4 \times \frac{22}{7} \times 14^2) + (2 \times \frac{22}{7} \times 14 \times 14)$ $\qquad = 1232 + 1232$ $\qquad = 2464$ b) Volume $= \pi r^2 h - \frac{1}{2} \times \frac{4}{3}\pi r^3$ $\qquad = (\frac{22}{7} \times 14^2 \times 14) - (\frac{2}{3} \times \frac{22}{7} \times 14^3)$ $\qquad = 8624 - 5749\frac{1}{3}$ $\qquad = 2874\frac{2}{3} cm^3$ $\qquad = (2874\frac{2}{3} \div 1000) \cong 2.875 litres$ c) Volume $= \frac{\pi r^2 h}{40}$ $\quad h = 5614 cm$ $\qquad = (\frac{22}{7} \times 14^2 \times 5614) \div 40 cm^3/sec$ $\qquad = 3458224 \div 40 cm^3/sec$ $\qquad = 86455.6$ seconds Time $\quad = \dfrac{86455.6}{3600} = 24.02 hrs$ $\qquad = 24hrs \ 1minute$
	Total

| 24 | a) | Sum of interior angles (2n - 4) x 90 |

a) Sum of interior angles (2n - 4) x 90
n – sides has n angles
Hence
$(2n - 4) \times 90^0 = 2(90^0) + 150^0(n - 2)$
$180n – 360^0 = 180^0 + 150^0n – 300^0$
$180n – 154n = 360^0 - 120^0$
$30n = 240$
n= 8 sides
$S = (2 \times 8 - 4) \times 90 = 1080^0$

b) Polygon - Heptagon

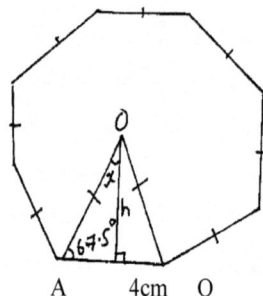

A   4cm   O

c) Angle at centre = $\dfrac{360^0}{8}$ = 450, base angle = ½ x 135 = 67.5⁰

x = ½ x 45⁰ = 22.5⁰,

$\tan 67.5^0 = \dfrac{h}{2}$

Therefore h = 2 tan 67.5⁰
h = 4.828cm
Angle of AOB = (½ x 4 x 4.828)
= 9.656
Area of octagon = 8 x area of 1 triangle
= 8 x 9.656
= 77.248

Total

CHAPTER EIGHT

1.　Without using logarithms table or calculator, solve for x in:

$$\log 5 - 2 + \log(2x + 10) = \log (x - 4)$$

2.　The initial salary of Mr. Kiptum is Sh. 42,000 per annum. His salary increases by 13% each year. Determine his total earnings after 15 years. Give your answer to the nearest thousands.

3.　Evaluate $\dfrac{Sin420^0 \; Cos330^0}{Tan570^0 \; Sin765^0}$ leaving your answer in surd form.

4.　In the figure below, O is the centre of the circle and TA and TB are tangents to the circle at A and B respectively. Given that angle ATO = 39^0; calculate angle TBX.

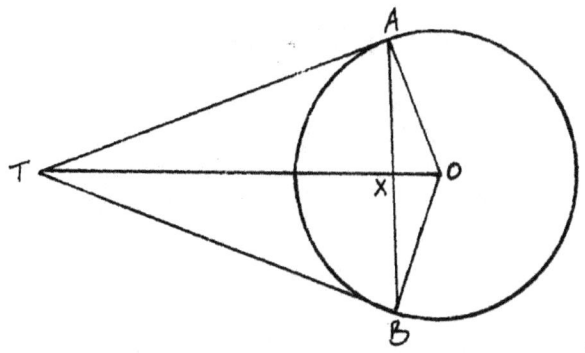

5.   Given the calculation 8.70 x 2.4, Rono obtains an approximate value by rounding off the given values to the nearest whole number. Determine the percentage error in the calculation arising from the approximation.

6.   For a lifting machine, the effort E required to lift a load L is partly constant and partly varies as L. Given that L = 2 when E = 5.5 and L = 6 when E = 6.5, Write an equation connecting E and L.

7.   Draw a line PQ = 7.2cm and on one side of the line, use a ruler and pair of compasses only to draw the locus of a point A such that $\angle$ PAQ = 60^0 and on it mark point A such that PA = QA.

8.   Mrs. Nandi bought a television set on hire purchase by paying a down payment of Ksh. 5000 and
Monthly installments of Ksh. 1250 for 2 years. If the interest rate charged was 12% p.a what is the carrying change to the nearest hundreds?

9.   A ball is dropped from the top of a building and its height h, metres above the ground at any time t, seconds is given by h = 350 + 65t – t².
    i)   Find the velocity of the ball when t = 2secs
    ii)   State the time when the ball hits the ground.

10.   Use matrix method to solve the simultaneous equations.
    $x + 3 (y - \frac{3}{4}) = \frac{1}{2}$
    $x = \frac{1}{3}y + \frac{1}{4}$
    $x = \frac{y}{3} + \frac{1}{4}$

11.   Atieno is now four times as old as her daughter and six times as old as her son. Twelve years from now, the sum of the ages of her daughter and son will differ from her age by 9 years. What is Atieno's present age?

12.   Solve for θ in the equation $Sin(3\theta + 120^0) = \frac{\sqrt{3}}{2}$ for $0 \le \theta \le 180^0$.

13.   A two digit number is such that the square of the unit digit is equal to one less than the tens digit and that the unit digit raised to power four and add three times the tens digit is equal to seven. Find the number.

14. i) Expand $\left(5+\dfrac{x}{2}\right)^6$ up to the term in x^3.

ii) Use your expansion to estimate the value of $\left(\dfrac{11}{2}\right)^6$. Correct to one decimal place.

15. A line segment joining two points P(0,7) and S (2, 3.8) is divided externally by point Q in the ratio 7:3. Find the co-ordinates of point Q.

16. The velocity of a particle, Vm/s moving in a straight line after t seconds is given by $V = 3t^2 - 3t - 6$. Find the distance covered by the particle between t = 1 and t = 4 seconds.

17. The probability that Hilda, Lucy and Caroline will be late for breakfast on any one morning are
¼ , $^1/_3$ and $^1/_5$ respectively on any one morning.
   a) Using a probability tree diagram find the probability that:-
      i) none of them will be late
      ii) Only one of them will be late
      iii) at least one of them will be late
      iv) at most one of them will be late

18. a) A figure whose co-ordinates are A(-2, -2), B(-4, -1), C(-4, -3) and D (-2, -3) undergoes
successive transformations ERS; where E, R and S are transformations represented by the matrices,
$$E=\begin{pmatrix} -2 & 0 \\ 0 & -2 \end{pmatrix}, \begin{pmatrix} 0 & -1 \\ -1 & 0 \end{pmatrix} \text{ and } R=\begin{pmatrix} 0 & 1 \\ -1 & 0 \end{pmatrix}$$
On the grid provided, show the figure ABCD and its image under the successive transformations ERS.

b) Find the matrix representing the single transformation mapping the image found in (a) above back the object figure ABCD.

c) Triangle PQR has vertices at P(2, 2), Q(4, 1) and R(6, 4). On the same grid, show the image of triangle PQR under a shear with line y = 2 invariant and point R(6, 4) is mapped onto R^1(2, 4).

19.   The positions of two towns on the earth's surface are A (40⁰S, 45⁰W) and B (40⁰S, 135⁰E)

   a)   Calculate;
The difference in distance between two towns A and B along the parallel of latitude and along the great circle. (in Nm).

b)   Two planes X and Y left town A at 8.00am flying at 758 knots each towards town B. If plane X flys along the parallel of latitude and plane Y along the great circle; then determine the position of one of the planes when the other lands at town B.

c)   What is the local time at town B when the second plane lands.

20.   The table below shows the distribution of wages in a week for a number of employees in a certain factory.

Wage	800 – 899	900 – 999	1000 – 1099	1100 – 1199	1200 – 1399	1400 – 15!
No. of employees	3	10	23	9	3	2

   a)   Using Kshs. 1049.50 per week as the assumed mean wage, calculate:-
      i) the mean for the group wages.
      ii) the standard deviation.

b)   The week that followed, every employee earned Ksh. 100 as wage increment.
            Determine:
         i)   the new mean for the group wage.
         ii) the new standard deviation

21.   The table below shows the corresponding values of X and Y which are connected by a relation of the form y = ab^x – 10, where a and b are constants.

X	1	2	3	4	5
Y	-6.5	4.1	47.5	197.5	947.5

      Draw a suitable line graph and find the appropriate numerical values of a and b. Use the grid below.

22.   The figure below is a solid in which the base ABCD is a rhombus. AC = 16cm, BD = 12cm and
         CE = 12cm.

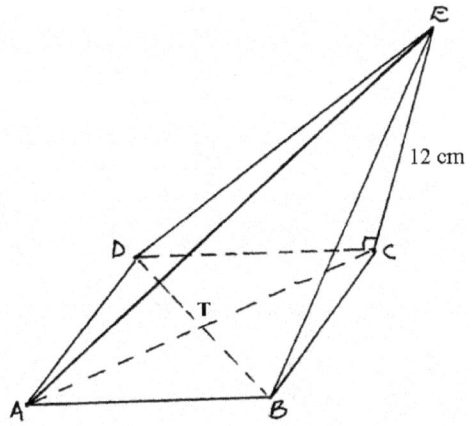

Calculate:-

   a)   the length of line BC

   b)   the angle between the planes EBD and ABCD

   c)   the angle between the planes ECB and ECD

   d)   the length of line AE

23.   The velocity V metres per second of a particle at time, t seconds is given by the equation below;

$$V = 2t^2 - 4t + 15$$

a)   Complete the table below for values of V and t

t	0	1	2	3	4	5	6	7	8
V	15								111

b)   On the grid below, draw the graph of V against t

c)   i)   Using the mid-ordinate rule with seven ordinates estimate the distance covered by the
        particle between t = 1 sec and t = 8sec.
        ii)   Determine the exact distance covered by the particle between t = 1 sec and t = 8 sec

        iii)   Find the percentage error in the distance covered by the particle when the mid-ordinate rule is used.

24.   Water is drawn to fill an empty tank whose capacity is 1200litres using two types of buckets. It requires at least 30 type A buckets and 50 type B buckets to fill the tank. Also, two type A buckets are required to fill at most three type B buckets. Each type B bucket has a capacity of not more than 20litres.
a)   Taking x litres and y litres to represent the capacity of each type A bucket and each type B bucket respectively; write down 3 inequalities to represent the above information.

b)   Use graphical method to determine the capacity of each type of bucket.

SOLUTIONS TO CHAPTER EIGHT

1	Log 5 – log 100 + log (2x + 10) = log (x - 4) $\text{Log}\left(\dfrac{x-4}{2x+10}\right)=\log\dfrac{5}{100}$ $\dfrac{x-4}{2x+10}=\dfrac{1}{20}$ 20x – 80 = 2x + 10 18x = 90 x = 5
	Total
2	Common ration = $\dfrac{113}{100}$ = 1.13 First term = Sh. 42, 000 $S_{15}=\dfrac{42{,}000\,(1.13^{15}-1)}{1.13-1}$ $=\dfrac{42{,}000\,(6.2543-1)}{1.13-1}$ $=\dfrac{42{,}000\,(6.2543-1)}{0.13}$ = 1, 697,533.50 $\cong$ 1,698,000
	Total
3	$\dfrac{\sqrt{3}/2 \times \sqrt{3}/2}{1/\sqrt{3} \times 1/\sqrt{2}}$ $= \dfrac{3}{4} \times \dfrac{\sqrt{6}}{1}$ $= \dfrac{\sqrt[3]{6}}{4}$
	Total
4	In ΔTBX, $\angle$BTX = 39^0 and $\angle$BXT = 90^0 Therefore; $\angle$TBX = 90^0 – 39^0 $\qquad\qquad = 51^0$
	Total
5	Approximation: 9 x 2 = 18 Actual value: 8.70 x 2.4 = 20.88 Absolute error: = 20.88 – 18 $\qquad\qquad$ = 2.88 $\qquad\qquad$ % error = $\dfrac{2.88}{20.88}$ x 100 $\qquad\qquad$ = 13.8%
	Total
6	E = K + CL where K and C are constants K + 2C = 5.5

	K + 6C = 6.5
	$\quad$ -4C= -1
	$\qquad$ C = ¼ ; K = 5
	Therefore; E = ¼ L + 5
	Total
7.
Locus of A
	Total
8	Hire purchase price = 5000 + (1250 x 24) = sh. 3500

$$P\left(1+\frac{12}{100}\times\frac{1}{12}\right)^{24} = 30000$$

$$P\left(1+0.01\right)^{24} = 30000$$

$$P=\frac{30000}{1.2697}$$

P = Sh. 23627.60
Carrying change = 35000 − 28627.60
$\quad$ P = Sh. 6372.40
$\quad$ ≅ Sh. 6400 |
| | Total |
| 9 | i) $\quad$ V = 65 – 2t
$\qquad$ = 65 – 4
$\qquad$ = 61m/s
ii) $\quad$ When h = 0
$\qquad$ Therefore t2 – 65t – 350 = 0
$\qquad$ (t + 5) (t-70) = 0
$\qquad$ t= -5 or 70
$\qquad$ Therefore time = 70secs
$\qquad\qquad$ Or 1min 10 secs |
| | Total |
| 10 | 4x + 12y = 11…….(i)
12x – 4y = 3………(ii) |

$$\begin{pmatrix} 4 & 12 \\ 12 & -4 \end{pmatrix} \begin{pmatrix} x \\ y \end{pmatrix} = \begin{pmatrix} 11 \\ 3 \end{pmatrix}$$

$$\begin{pmatrix} x \\ y \end{pmatrix} = \frac{1}{-160} \begin{pmatrix} -4 & -12 \\ -12 & 4 \end{pmatrix} \begin{pmatrix} 11 \\ 3 \end{pmatrix}$$

$$\begin{pmatrix} x \\ y \end{pmatrix} = \begin{pmatrix} \frac{1}{2} \\ \frac{3}{4} \end{pmatrix}$$

Therefore $x = \frac{1}{2}$, $y = \frac{3}{4}$

		Total
11	Let Atieno's present age be x yrs. Then daughter's age = $\frac{x}{4}$ Son's age = $\frac{x}{6}$ In 12yrs time; Atieno's age = $(x + 12)$ Daughter's age = $(\frac{x}{4} + 12)$ Son's age = $(\frac{x}{6} + 12)$ Therefore $(\frac{x}{4} + 12) + (\frac{x}{6} + 12) = (x + 12) - 9$ $\frac{3x + 2x}{12} = x + 12 - 9 - 24$ $\frac{5x}{12} = x - 21$ $5x = 12x - 252$ $-7x = -252$ $x = 36yrs$	
		Total
12	$\sin 60^0 = \frac{\sqrt{3}}{2}$ Therefore $3\theta + 120^0 = 60^0, 120^0, 420^0, 480^0$ $3\theta = -60^0, 0^0, 300^0, 360^0$ $\theta = -20^0, 0^0, 100^0, 120^0$ Hence $\theta = 0^0, 100^0,$ or 120^0	
		Total
13	Let the number be xy; Then $y^2 = x - 1$ ..... (i) $Y^4 = 3x = 7$ ....(ii) $(x-1)^2 + 3x = 7$ $x^2 + x - 6 = 0$ $(x-2)(x+3) = 0$ $x = 2$ or -3 $x = 2, y = 1$ the number is 21	
		Total

14	i) Coefficients: 1 6 15 20
	    Terms: 5^6, $5^5 (\frac{x}{2})$, $5^4 (\frac{x}{2})^2$, $5^3 (\frac{x}{2})^3$
	    $15625, \dfrac{3125x}{2}, \dfrac{625x^2}{4}, \dfrac{125x^3}{8}$	

$$\left(5+\frac{x}{2}\right)^6 = 15625 + \frac{3125}{3}x + \frac{9375x^3}{4} + \frac{625x^3}{2} + \ldots$$

ii) $x=1$ *Therefore,* $\left(\frac{1}{2}\right) = 15625 + \frac{3125}{3} + \frac{9375}{4} + \frac{625}{2}$

$\quad = 15625 + 1041.667 + 2343.75 + 312.5$

$\quad = 19322.9$

		Total

15 PQ : QS = 7: -3

Ratio theorem : $OQ = \frac{7}{4}OP + \frac{-3}{4}OS$

$OQ = \frac{7}{4}\begin{pmatrix} 0 \\ 7 \end{pmatrix} + \frac{-3}{4}\begin{pmatrix} 2 \\ 3.8 \end{pmatrix}$

$= \begin{pmatrix} 0 \\ \frac{49}{4} \end{pmatrix} - \begin{pmatrix} \frac{3}{2} \\ \frac{11.4}{4} \end{pmatrix}$

$= \begin{pmatrix} \frac{-3}{2} \\ 9.4 \end{pmatrix}$

Q is at (−1.5, 9.4)

		Total

16

$$Dis\tan ce\,cov ered = \int_1^4 3t^2 - 3t - 6dt$$

$$= \left[t^3 - \frac{3t^3}{2} - 6t \right]_1^4$$

$$= \left(4^3 - \frac{3(4)^2}{2} - 6(4) \right) - \left(1^3 - \frac{3(1)^2}{2} - 6(1) \right)$$

$$= 16 - \left(\frac{-13}{2}\right)$$

$$= 22\frac{1}{2}\,m\;or\;22.5m$$

		Total

17

a) H – Hilda C – Caroline L- Lucy

b) i) $P(H^1C^1L^1) = \frac{3}{4} \times \frac{2}{3} \times \frac{4}{5}$
$= \frac{2}{5}$ or 0.4

ii) $P(H^1C^1L^1$ or $H^1C^1L^1$ or $H^1C^1L^1)$
$= (\frac{1}{4} \times \frac{2}{3} \times \frac{4}{5}) + (\frac{3}{4} \times \frac{1}{3} \times \frac{4}{5}) + (\frac{3}{4} \times \frac{2}{5} \times \frac{1}{5})$
$= \frac{13}{30}$ or 0.4333

iii) $1 - P$ (none will be late)
$= 1 - \frac{3}{5}$
$= \frac{3}{5}$

iv) P (None will be late or only one is late)
$= \frac{2}{5} + \frac{13}{30}$
$= \frac{5}{6}$

		Total

18

$$\begin{pmatrix} 0 & -1 \\ -1 & 0 \end{pmatrix} \begin{pmatrix} A & B & C & D \\ -2 & -4 & -4 & -2 \\ -2 & -1 & -3 & -3 \end{pmatrix} = \begin{pmatrix} A' & B' & C' & D' \\ 2 & 1 & 3 & 3 \\ 2 & 4 & 4 & 2 \end{pmatrix}$$

a)
$$\begin{pmatrix} 0 & 1 \\ -1 & 0 \end{pmatrix} \begin{pmatrix} A' & B' & C' & D' \\ 2 & 1 & 3 & 3 \\ 2 & 4 & 4 & 2 \end{pmatrix} = \begin{pmatrix} A'' & B'' & C'' & D'' \\ 2 & 4 & 4 & 2 \\ -2 & -1 & -3 & -3 \end{pmatrix}$$

$$\begin{pmatrix} -2 & 0 \\ 0 & -2 \end{pmatrix} \begin{pmatrix} A'' & B'' & C'' & D'' \\ 2 & 4 & 4 & 2 \\ -2 & -1 & -3 & -3 \end{pmatrix} = \begin{pmatrix} A''' & B''' & C''' & D''' \\ -4 & -8 & -8 & -4 \\ 4 & 2 & 6 & 6 \end{pmatrix}$$

b) $S^1 R^1 E^1 (A'''B'''C'''D''') = ABCD$

$$\begin{pmatrix} 0 & -1 \\ -1 & 0 \end{pmatrix} \begin{pmatrix} 0 & -1 \\ 1 & 0 \end{pmatrix} \begin{pmatrix} -\frac{1}{2} & 0 \\ 0 & -\frac{1}{2} \end{pmatrix}$$

$$= \begin{pmatrix} 0 & -1 \\ -1 & 0 \end{pmatrix} \begin{pmatrix} 0 & -\frac{1}{2} \\ -\frac{1}{2} & 0 \end{pmatrix}$$

$$= \begin{pmatrix} \frac{1}{2} & -1 \\ 0 & -\frac{1}{2} \end{pmatrix}$$

c) ΔPQR – For object figure plotted correctly

PQR – For image figure plotted correctly

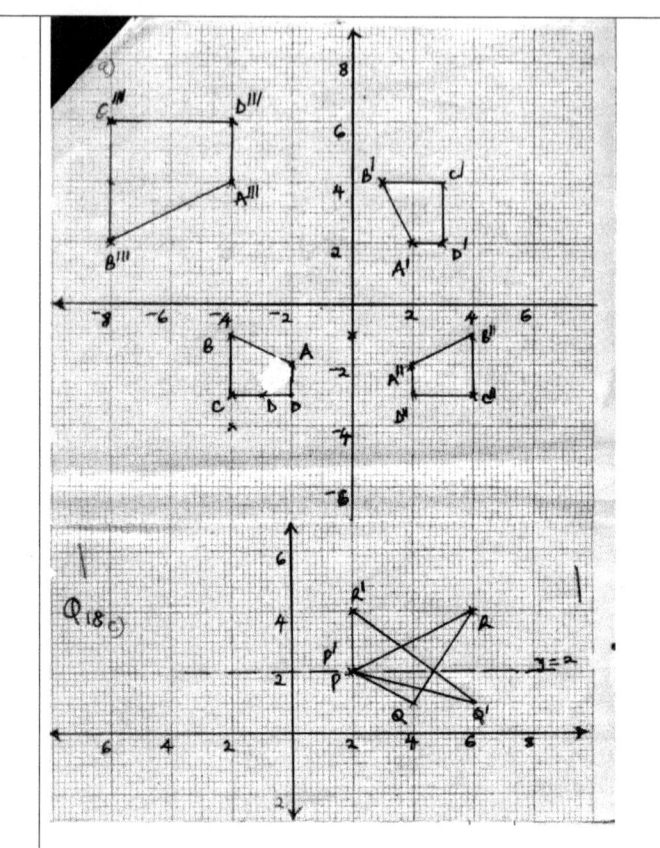

	Total

19 | Along parallel latitude.
a) $180 \times 60 \cos 40 = 8273.28$nm
   Along the great circle.
   $100 \times 60 = 6000$nm
   Difference = 2273.28nm

b) Time taken by $x = \dfrac{8273.28}{758}$

   10hrs 55min

   Time taken by $y = \dfrac{6000}{758}$

   7hrs 55min
   Time difference = 3hrs
   Distance to cover = $3 \times 758 = 2274$

   $Q = \dfrac{2274}{60\cos 40} = 49.50^0$

   Therefore $135 - 49.5 = 85.5^0$
   $(40^0S, 85.5^0E)$

c) 8.00 am + 10.55 = 6.55 pm
+ 12hour time difference 6.55am next day.

Total

20.

	x	f	t	$^{t}/_{50}$	$^{ft}/_{50}$	$(^{t}/_{50})^2$	$(^{ft}/_{50})^2$
800-899	849.5	3	-200	-4	-12	16	48
900-999	949.5	10	-100	-2	-20	4	40
1000-099	1049.5	23	0	0	0	0	0
1100-1199	1149.5	9	100	2	18	4	36
1200-1399	1299.5	3	250	5	15	25	75
1400-1599	1499.5	2	450	9	18	81	162
					19		361

Mean = $1049.5 + (^{19}/_{50})$

= 1068.5

$$S.d = 50\sqrt{\frac{361}{50} - \left(\frac{19}{50}\right)^2}$$

$$= 50\sqrt{7.22^2 - 0.1444}$$

50 x 2.66

= 133

b) New mean 1068.5 + 100 = 1168.5

S.d = 133

Total

21.

$y+10 = ab^x$

$\log(y+10) = x \log b + \log a$

X	1	2	3	4	5
y+10	3.5	14.1	57.5	207.5	957.5
log(y+10)	1.544	1.1492	1.7597	2.317	2.9811

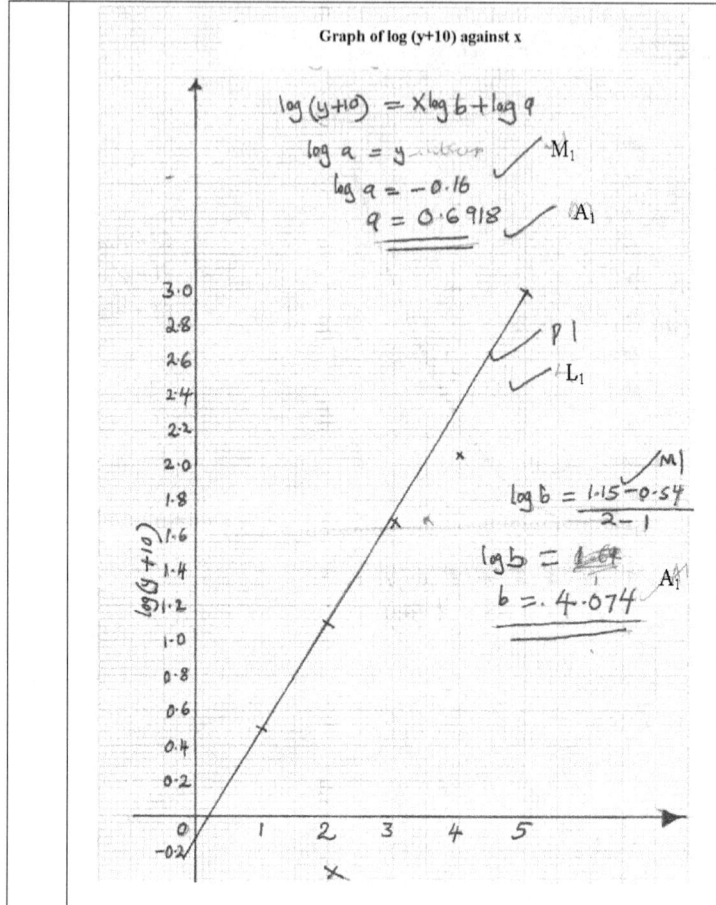

Graph of log (y+10) against x

$\log (y+10) = x \log b + \log a$

$\log a = y \text{ intercept}$

$\log a = -0.16$

$a = 0.6918$ ✓ A₁

$\log b = \dfrac{1.15 - 0.54}{2 - 1}$

$\log b = \cancel{1.64}$ A₁

$b = 4.074$ ✓

		Total

22.	a) length of BC

$$BC = \sqrt{8^2 + 6^2}$$

$$= \sqrt{100} = 10cm$$

b) TC = 8cm   EC = 12cm

$$Tan\,\theta = \frac{12}{8} = 1.5$$

$$\theta = 56.31^0$$

c) Tan $\theta = {}^6/_8$ $\theta = 36.87^0$

Angle required = $36.87^0 \times 2 = 73.74^0$

d) Length of line AE = $\sqrt{16^2 + 12^2}$

$$= \sqrt{256 + 144}$$

$$= 20$$

136

						Total			

23 a)

T	0	1	2	3	4	5	6	7	8
V	15	13	15	21	31	45	63	85	111

b)

Graph of V against t.

SI –√s
PI – √
CI – V

c) i) Distance = 1(13.5+17.5+25.5+37+5.53.5+73.97.5)

         = 318.5 metres

   ii) Distance = $\int_{1}^{8} 2t^2 - 4t + 15 dt = \left[\frac{2}{3}t^3\ 2t^2 + 15t\right]_{1}^{8}$

         = 318.5metres

	iii) Percentage error $= \dfrac{319.67 - 318.5}{319.6} \times 100$ $= 0.3649\%$
	Total

24	a) Three inequalities:-
	      $30x + 50y \geq 1200...$
	      $3x + 5y \geq 120.....(i)$
	      $2x \leq 3y .... (ii)$
	      $y \leq 20....(iii)$
	 b) Equations:
	      $3x + 5y = 120...(i)$
	      $y = {}^2/_3 x....(ii)$
	      $y = 20....(iii)$
	 Search line: $30x + 50y = K$
	 Using point (20, 20), $K = 600 + 1000 = 1600$
	
	 Therefore, search line: $30x + 50y = 1600$
	                             $3x + 5y = 160$
	 The number of: - type A buckets – 19
	 type B buckets - 13

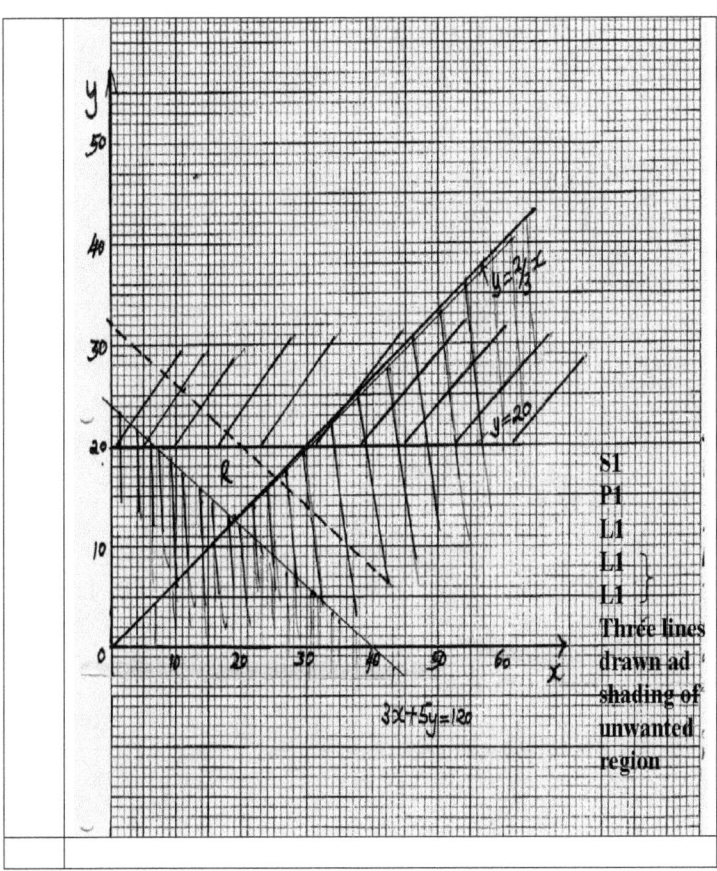

S1
P1
L1
L1 ⎫
L1 ⎬
Three lines
drawn ad
shading of
unwanted
region

CHAPTER NINE

1. **Evaluate:** $\dfrac{4 \times 6 + \dfrac{1}{25} \div 0.05 + \dfrac{1}{5}}{(-3) \div (-6) + (23) - 6 \ of \ 3}$

2. **Simplify** the expression $\dfrac{2x^2 - 3xy - 2y^2}{4x^2 - y^2} \div \dfrac{2x + y}{2x - y}$

3. The price of foodstuff generally increased by 20% at the beginning of a drought season and reduced by 30% during harvesting season. **Express** the new price at a ratio of the original price in its lower form.

4. Find the integral values of x which satisfy the inequalities $\begin{array}{c} 15 - 2x > 4 \\ 4 < 3x - 2 \end{array}$

5. A circle of radius 15cm is divided into ten equal sectors. In each sector, **find**:

    (a) The area of the triangle

    (b) The area of the segment

6. Tap A fills a water tank in 30min, B in 20mins and C in 10mins. All three taps are turned on from 8:55am and then C is turned off. At what time will the tank be filled after C has been closed?

7. The logarithms of the squares of a and b are 1.204 and 0.954 respectively. **Find** the logarithms of their product.

8. The mean of a set of n numbers is 28. If an extra number 18 is included in the set, the mean now becomes 26. **Find** the value of n

9. In the figure below, $\angle EHG = \angle EFH = 90^0$ HF = 5cm, and EF = 12cm. calculate the lengths HG and FG.

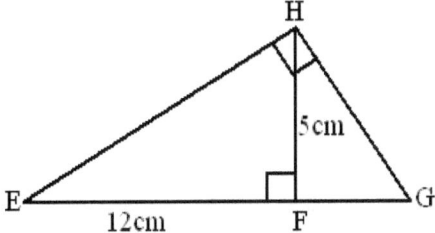

10. The line $y = mx + 6$ makes an angle of $75^0\ 58'$ with x – axis. **Find** the coordinates of the point where the line cuts the x-axis.

11. Find the equation of the image of the line $y = 3x + 5$ under reflection in the line $x = y$.

12. Given that log y = 3.143 and log x = 2.421, **evaluate:** $4 \log y^{\frac{1}{2}} + \log \sqrt[3]{x}$

13. (i) **Express** 98 and 72 as products of their prime factors.

(ii) A rectangle of side 98cm by 72cm is subdivided into small squares each of side x cm.
Find the values of x.

14. The co-ordinates of points A, B and C are (0, -4), (2, -1) and (4, 2) respectively. **Use** vectors to show that the points A, B and C are collinear.

15. If $2^{x+y} = 16$ *and* $4^{2x-y} = \dfrac{1}{4}$,   find the ratio y – x : 2y

16. Determine the lower quartile, upper quartile and the quartile deviation for the following set of numbers. 5, 10, 6, 5, 8, 7, 3, 2, 7, 8, 9.

17. The following are masses of 25 students in form 4 class.

49, 51, 50, 60, 55

45, 56, 51, 58, 59

44, 42, 59, 50, 62

46, 43, 57, 56, 52

43, 41, 40, 54, 44

(a) Draw a frequency distribution table with the lower class 40 – 43

(b) Estimate the median mass

© **Draw** a histogram for the data.

18. In the figure below O is the centre and PS is a diameter of the circle. QR is parallel to PS. If angle PSQ is 25^0 and angle POT is 120^0, **find** the sizes of the given angles giving reasons.

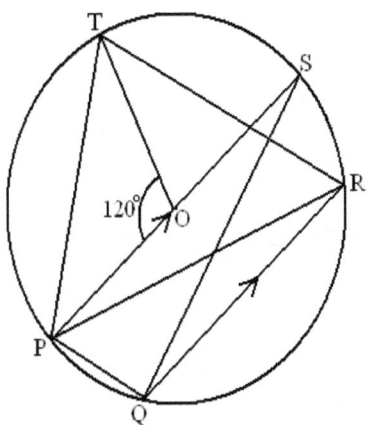

Angle QRT

Angle QPT

Angle PQR

Angle PTR

19. A bus left Nairobi at 7.00a.m and travelled towards Eldoret at an average speed of 80Km/hr. At 7.45a.m a car left Eldoret towards Nairobi at an average speed of 120Km/hr. the distance between Nairobi and Eldoret is 300km. **Calculate**:

(a) The time the bus arrived at Eldoret.

b.The time of the day, the two met.

c. The distance from Nairobi where the two met.

d. The distance of the bus from Eldoret when the car arrived at Nairobi.

20. A three digit number is such that the sum of its hundreds and tens digits is 10. When the number is divided by its hundreds digit, the quotient is 108. If the number is divided by the sum of all the digits, the quotient is 36. **Find** the number

21.   The figure below represents the cross-section of a tunnel. The cross-section is in the form of a major segment of a circle. M is the mid-point of AB and CM is perpendicular to AB. Given that AB = CM = 8cm, **Calculate** the volume of the tunnel if it is 15cm long.

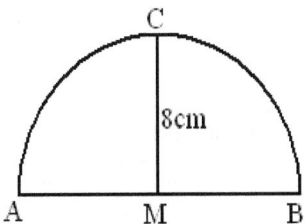

8cm

C

A   M   B

22. In the figure below C is a point on AB such that $BA = 3BC$ and D is the mid-point of

OA. OC and BD intersect at xGiven that $OA = a$ and $OB = b$

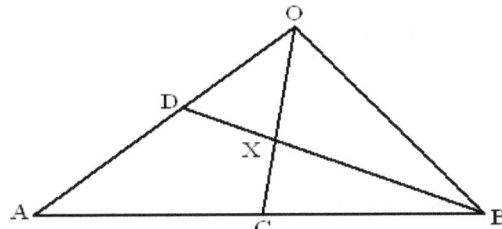

(a) **Write** down in terms of a and b the vectors.

(i) AB

(ii) OC

(iii) BD

(b) If BX = h.BD, express OX in terms of a , b and h

© If $OX = kOC$, find h and k

145

d. Hence express OX in terms of a and b only.

23. (a) Complete the table below, giving your values correct to 2 decimal places.

X	0	15	30	45	60	75	90	105	120	135	150	165	180
Cos 2x	1.0		0.5	0		- 0.87	-1.0		-0.5		0.5	0.87	0
Cos (2x + 30)		0.5	0	-0.5		-1.0		-0.5	0		0.87	1.0	0.87

b. Using the grid provided, on the same axes **draw** the graphs of y = Cos 2x and y = Cos (2x + 30⁰) Use the scale 1cm for 15⁰ on the x – axis, 5cm for 1 unit on y-axis.

c. **State** the amplitude of each graph.

d. Use your graph to **determine**:

(i) The solution to the equation: Cos (2x + 30) – Cos 2x = 0.

(ii) The transformation that would map the graph of y = Cos 2x onto the graph of

y = Cos (2x + 30)

24. (a) Three villages A, B and C are such that B is 3km on a bearing of 030⁰ from A, C is 4km on a   bearing of 120km from B.

(i) Using a scale of 1cm to represent 0.5km, **draw** a diagram to show the relative positions of the village A, B and C.

(ii) **Find** the distance and bearing of village A from C.

(iii)A straight main road runs from village A to C. **Find** the length of the shortest path from village B to the main road.

(b) The measurements (in metres) of a field were given in a field note book as follows:

Base line XY = 240m

	Y	
TO R 60	190	
	150	50 TO P
TO Q 60	120	
TO T 30	50	20 TO M
	X	

(i) **Make** a sketch of the field

(ii) **Find** the area of the field in hectares.

SOLUTIONS TO CHAPTER NINE

NO.	WORKING
1.	$$\frac{24 + 0.8 + 0.2}{0.5 + 23 - 18}$$ $$= \frac{25}{5.5} = \frac{25}{55} \times 10 = \frac{50}{11}$$ $$= 4\,{}^{6}\!/_{11} \; OR$$ $$= 4.545$$
2.	$$\frac{2x^2 - 4xy + xy - 2y^2}{(2x - y)(2x + y)} \div \frac{2x + y}{2x - y}$$ $$= 2x\frac{(x - 2y) + y\,(x - 2y)}{(2x - y)(2x + y)} \times \frac{(2x - y)}{2x + y}$$ $$= \frac{(2x + y)(x - 2y)}{(2x - y)(2x + y)} \times \frac{(2x - y)}{(2x + y)}$$ $$= \frac{x - 2y}{2x + y}$$
3.	Let original price be x Then after 20% increase becomes 1.2x After 30% decrease Becomes $\left(\dfrac{70}{100} \times 1.2x\right)$ $= 0.84x.$ $\therefore$ ratio 0.84x:x $= 84{:}100$ $= 21{:}25$

4.	$2x < 15 - 4$ ....(i) $x < 5.5$ $3x > 4 + 2$ ....(ii) $3x > 6$ $x > 2$ $\therefore 2 < x < 5.5$ int *ergral values* $\{3, 4, 5\}$
5.	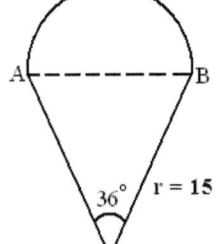 $\dfrac{360}{10} = 36^0$ (a) *Area of* $\triangle$ *OAB* $= \dfrac{1}{2}.15.15 Sin36$ $\quad = \dfrac{1}{2}.225 \times 0.5878$ $\quad = 66.13cm^2 \ (4s.f)$ (b) *area of segment* $\quad = Area \ of \ \sec tor - Area \ of \ \triangle$ $\quad = \dfrac{36}{360} \times \pi r^2 - \dfrac{1}{2}B^2 \ Sin36$ $\quad = \dfrac{1}{10}.\dfrac{22}{7}.15.15 - \dfrac{1}{2}.15.15.0.5878$ $\quad = 70.71 - 66.13$ $\quad = 4.58cm^2$
6.	For all taps together fills

$$\left(\frac{1}{30} + \frac{1}{20} + \frac{1}{10}\right) \; of \tan k \; in \; 1 \; min$$

$$= \frac{1}{60} \; of \; \tan k$$

$\therefore in \; 4 \; min \; (8.59 - 8.55)$

$$= \left(4 \times \frac{11}{60}\right) \; of \; \tan k \; filled$$

$$= \frac{11}{15} \; of \; \tan k \; filled$$

$Re\,maining \; portion \; is \; 1 - \dfrac{11}{15} = \dfrac{4}{15} \; of \; the \; \tan k$

$for \; A \; \& \; B \; fills \; \dfrac{5}{60} \; of \; \tan k \; in \; 1 \; min$

$\therefore \dfrac{4}{15} \; of \; \tan k \Rightarrow ?$

$$\left({}^{4}\!/_{15} \times 1 \times \frac{\overset{4}{\cancel{60}}}{5}\right) \min s$$

$= \dfrac{16}{5} \; min = 3\,min \; 12 \; sec$

$time = 8.59 + 3\,min \; 12 \; sec$

$\qquad = 9 : 02 : 12 \; am$

7.	$\log a^2 = 1.204$ $\therefore \log a = \dfrac{1.204}{2}$ $\qquad = 0.602$ $\log b^2 = 0.954$ $\therefore \log b = 0.477$ $\therefore \log (a \times b) = \log a + \log b$ $\qquad\qquad = 0.602 + 0.477$ $\qquad\qquad = 1.079$
8.	Sum of n numbers = 28n Sum of n+1 numbers $\qquad = 26\,(n+1)$

	$= 26n + 26$ $\therefore 28n + 18 = 26n + 26$ $2n = 8$ $n = 4$
9.	DEHG III DEFG $\Rightarrow \dfrac{EH}{EF} = \dfrac{HG}{FH} = \dfrac{EG}{EG}$ $\therefore \dfrac{13}{12} = \dfrac{HG}{5}$ $\Rightarrow HG = \dfrac{13 \times 5}{12} = \dfrac{65}{12}$ $\qquad = 5.417 \ (4sf)$ $\therefore FG = \sqrt{\left(\dfrac{65}{12}\right)^2 - 5^2}$ $\qquad = \sqrt{(29.34 - 25)}$ $FG = \sqrt{\dfrac{625}{144}} = \dfrac{25}{12}$ $\qquad = \sqrt{4.34}$ $\qquad = 2.083 \ cm$
10.	

	$M = Gradient = \tan 75^0\ 58^1$
	$= 4.001$
	$Pts\ (0, 6),\ (x, 0)$
	$\dfrac{0 - 6}{x - 0} = 4$
	$-6 = 4x$
	$\therefore x = \dfrac{-6}{4} = -1.5$
	$\therefore x - intercept\ (-1.5, 0)$

11.	
	<table><tr><td>Pts on line</td><td>(a, b)</td><td>(0, 5)</td><td>(-1, 2)</td></tr><tr><td>Pts on image</td><td>(b, a)</td><td>(5, 0)</td><td>(2, -1)</td></tr></table>
	$\therefore$ Gradient of image $= \dfrac{-1 - 0}{2 - 5}$
	$= \dfrac{1}{3}$
	equation of image $\Rightarrow \dfrac{y - 0}{x - 5} = \dfrac{1}{3}$
	$y = \dfrac{1}{3}(x - 5)$

12.	
	$2\log y + \dfrac{1}{3}\log x$
	$= 2\,(3.143) + \dfrac{1}{3}\,(2.421)$
	$= 6.286 + 0.807$
	$= 7.093$

13.	
	$(i)\ \ 98 = 2 \times 7^2$
	$72 = 2^3 \times 3^2$
	$(ii)\ (98 \times 72)\,cm^2 = 7056$
	$GCD\ of\ 98\ \&\ 72 = 2$
	$\therefore x = 2cm$

14.	$OA = \begin{pmatrix} 0 \\ -4 \end{pmatrix}$, $OB \begin{pmatrix} 2 \\ -1 \end{pmatrix}$ $OC = \begin{pmatrix} 4 \\ 2 \end{pmatrix}$ $AB = OB - OA = \begin{pmatrix} 2 \\ -1 \end{pmatrix} - \begin{pmatrix} 0 \\ -4 \end{pmatrix}$ $= \begin{pmatrix} 2 \\ 3 \end{pmatrix}$ $BC = OC - OB = \begin{pmatrix} 4 \\ 2 \end{pmatrix} - \begin{pmatrix} 2 \\ -1 \end{pmatrix}$ $= \begin{pmatrix} 2 \\ 3 \end{pmatrix}$ $\therefore\ AB = BC$, *but B is a common pt* $\therefore\ A, B, C$ *are collinear.*
15.	$2^{x+y} = 2^4 \Rightarrow x + y = 4 ....(i)$ $4^{2x-y} = 4^{-1} \Rightarrow 2x - = -1 ....(ii)$ *Solving simul* tan *eously* $x + y = 4$ $2x - y = -\overset{+}{1}$ $3x = \quad 3$ $x = 1, \Rightarrow y = 3$ $y - x = 3 - 1 = 2$ $2y = 2 \times 3 = 6$ $\therefore\ y - x : 2y$ $= 2 : 6$ $= 1 : 3$
16.	Arrange data in ascending order. 2, 3, 5, 5, 6, 7, 7, 8, 8, 9, 10 median = 7 lower quartile = 5 upper quartile = 8 quartile deviation = semi – Interquartile range $= \dfrac{8 - 5}{2} = \dfrac{3}{2}$ $= 1.5$

NO.	WORKING

17. (a)

class	mid point x	f	cf
40 – 43 ✔	41.5	5	5
39.5 – 43.5			
44 -47 ✔	45.5	4	9
43.5 – 47.5			
48 - 51 ✔	49.5	5	14
47.5 – 51.5			
52 - 55 ✔	53.5	3	17
51.5 – 55.5			
56 – 59 ✔	57.5	6	23
55.5 – 59.5			
60 - 63 ✔	61.5	2	25
59.5 – 63.5		✔	

(b) Median = $(12^{th} + 1)$th Student

$\qquad$ = 13th student

mass of 13th student

$$= 47.5 + \frac{\left(\frac{26}{2} - 9\right)4}{5}$$

$$= 47.5 + \frac{4 \times 4}{5}$$

$$= 47.5 + 3.2$$

$$= 50.7$$

18.

$$\angle PRT = \frac{1}{2} \angle POT \text{ ....}$$

$\qquad\qquad\qquad\qquad$ subtended at circumference & centre

respectively by chord PT

$= \frac{1}{2} \cdot 120$

$= 60^0$

but $\angle PRS = 90^0$

$\therefore \angle TRS = 30^0$

$\angle PST = 60^0 = \angle PRT$

Subtended by same chord PT

$\angle QRP = 25^0 \ldots$

$\angle PRT = \frac{1}{2} \times 120 = 60^0$

$\therefore \angle QRT = 25 + 60$

$= 85^0$

$\angle QPT = 180 - 85$

95^0

$\angle RQS = 25^0$ alt to $\angle PSQ$

$\angle PQS = 90^0 \ldots$ subtended by diameter

$\therefore PQR = 90 + 25$

$= 115^0$

$\angle PQR = 115^0$

$\angle PTR = 180 - 115^0$

$= 65^0 \ldots \ldots$ opps $\angle S$ of cyclic quadrilateral

19.	

(a) $time = \frac{dis \tan ce}{speed}$

$= \frac{300}{80}$

$= 3hrs \; 45\min$

$Arrival \; time = 7.00 + 3hr \; 45\min$

$= 10.45 a.m$

$(b) Re lative \; speed = 120 + 80 = 200 km / hr$

$Dis \tan ce \; separating \; them \; at \; 7.45 \; a.m$

$= 300 - \left(\frac{45}{100} \times 80 \right) = 240 km$

$t = \frac{240}{200} = 1hr \; 12\min$

$meeting \; time = 7.45 + 1hr \; 12 \; \min$

$= 8.57 am$

	$(c)\ D = \left(\dfrac{240}{200} \times 80\right) + \left(\dfrac{45}{60} \times 80\right)$ $= 156\ km$
	(d) *Arrival time for the car* $= 7.45 + \dfrac{300}{120}$ $= 10:15\ am$ $D = \left(\dfrac{10:45 - 10.15}{60} \times 80\right)$ $D = 8 \times 5$ $= 40km$
20.	Let the number be XYZ $= 100x + 10y + z$ $x + y = 10 \ldots\ldots\ldots(i)$ $\dfrac{100x + 10y + z}{x} = 108 = \ldots..(ii)$ $8x - 10y - z = 0 \qquad (ii)$ $\dfrac{100x + 10y + z}{x + y + z} = 36 \quad (iii)$ $64x - 26y - 35z = 0 \quad (iii)$ *from equation* $(ii)\ z = (8x - 10y)$ *substituting this* int$o$ (iii) *becomes* $64x - 26y - 35\,(8x - 10y) = 0$ $\Rightarrow 324y - 216x = 0$ $81y - 54x = 0 \ldots\ldots(iv)$ *multiplying* (i) *by* 54 *and* $u$sin$g$ $e$lim*ination becomes* $81y - 54x = 0\ldots\ldots\ldots(iv)$ $54y + 54x = 540$ $135y \qquad = 540$ $y = 4$ $\therefore x = 6$ $z = (48 - 40)$ $= 8$ $\therefore$ *The number is* 648

21.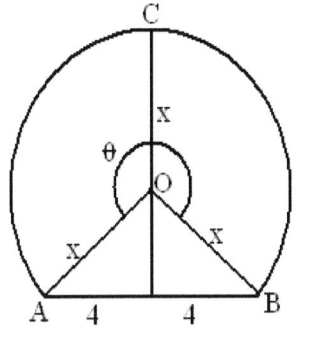

Let x be radius of circle

$\therefore (8-x)^2 + 4^2 = x^2$

$64 - 16x + x^2 + 16 = x^2$

$-16x = -80$

$x = 5cm$

Area of triangle AOB

= 1/2 .8.3

= 12cm^2

Area of sector ACB

$= \dfrac{\theta}{360} \times \pi r^2$

$\sin^{-1}(0.8) = \alpha$

$\alpha = 53.13$

$\therefore \theta = 360 - (2 \times 53.13)$

= 253.74

$\therefore$ Area of sector ACB

$= \dfrac{253.74}{360} \times \dfrac{22}{7} \times 5 \times 5$

= 55.38cm^2

Area of segment = Area of AOB + Area of sector ACB

$= (12 + 53.38)cm^2$

$= 67.38 \ cm^2$

Area of tunnel = length x cross – sectional area

= 15 x 67.38

= 1010.7cm^3

22.	

(a) (i) $AB = OB - OA$

$$= \underset{\sim}{b} - \underset{\sim}{a}$$

(ii) $OC = OA + \dfrac{1}{2} AB$

$$= \underset{\sim}{a} + \dfrac{1}{2}\left(\underset{\sim}{b} - \underset{\sim}{a}\right)$$

$$= \dfrac{1}{2}\underset{\sim}{a} + \dfrac{1}{2}\underset{\sim}{b}$$

(iii) $BD = -OB + \dfrac{1}{2}OA$

$$= -\underset{\sim}{b} + \dfrac{1}{2}\underset{\sim}{a}$$

(b) $Bx = h.\, BD$

$$= h\left(\dfrac{1}{2}\underset{\sim}{a} - \underset{\sim}{b}\right)$$

but $Bx = Ox - OB$

$\therefore OX = h..\, BD + OB$

$$= h\left(\dfrac{1}{2}\underset{\sim}{a} - \underset{\sim}{b}\right) + \underset{\sim}{b}$$

$$= \dfrac{h}{2}\underset{\sim}{a} - h\underset{\sim}{b} + \underset{\sim}{b}$$

(c) $OX = k.\left(\dfrac{1}{2}\underset{\sim}{a} + \dfrac{1}{2}\underset{\sim}{b}\right)$

but $OX = \dfrac{h}{2}\underset{\sim}{a} - h\underset{\sim}{b} + \underset{\sim}{b}$

as in (b) above.

$\therefore \dfrac{h}{2}\underset{\sim}{a} \, h\underset{\sim}{b} + \underset{\sim}{b} = \dfrac{k}{2}\underset{\sim}{a} + \dfrac{k}{2}\underset{\sim}{b}$

$\left(\dfrac{h}{2} - \dfrac{k}{2}\right)\underset{\sim}{a} = \left(\dfrac{k}{2} + h - 1\right)\underset{\sim}{b}$

collecting like vectors

	since $a \not\!\chi b$
	then
	$\dfrac{h}{2} - \dfrac{k}{2} = 0 \ \& \ \dfrac{k}{2} + h - 1 = 0$
	$h - k = 0 \ \& \ k + 2h = 2$
	solving simulteneously
	$h - k = 0$
	$+ \ 2h + k = 2 \ ^{+}$
	$\qquad 3h = 2$
	$h = \dfrac{2}{3}$
	$k = \dfrac{2}{3}$
	$(d) \ \underset{\sim}{OX} = k\left(\dfrac{1}{2}\underset{\sim}{a} + \dfrac{1}{2}\underset{\sim}{b}\right)$
	$\qquad = \dfrac{2}{3}\left(\dfrac{1}{2}\underset{\sim}{a} + \dfrac{1}{2}\underset{\sim}{b}\right)$
	$\qquad = \dfrac{1}{3}\underset{\sim}{a} + \dfrac{1}{3}\underset{\sim}{b}$
	$\qquad = \dfrac{1}{3}\left(\underset{\sim}{a} + \underset{\sim}{b}\right)$
23.	See the graph
24.a	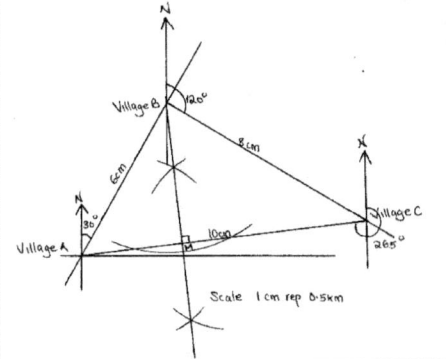

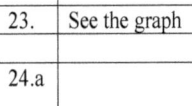

(ii) Distance from A – C

$$is \left(\frac{10 \times 0.5}{1}\right) km$$

$$= 5km$$

bearing of village A from C

is 265^0

(iii) Drop a perpendicular to AC through B & measure ie. BM

= 5cm

$$\therefore length\ of\ shortest\ path = \left(5 \times 0.5\right)km$$

$$= 2.5\,km$$

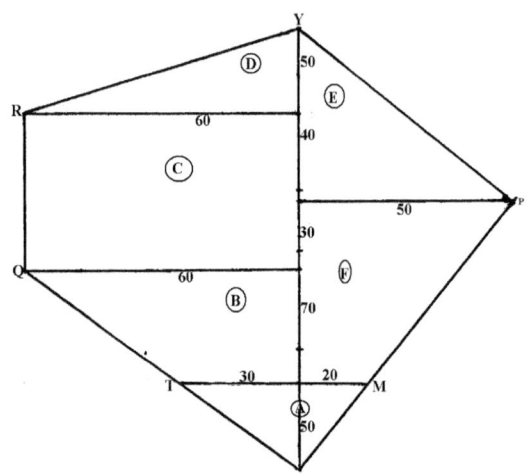

Area of A $= ½ .50.50 = 1250m^2$
Area of B $= ½ (30 + 60)x70 = 3150m^2$
Area of C $= 60 \times 70 = 4200m^2$
Area of D $= 1/2 .60.50 = 1500m^2$
Area of E $= ½ .90.50 = 2250m^2$
Area of F $= ½ (20 + 50) \times 100 = 3500m^2$
Total area $= 15,850m^2$
Area in hectares $= 1.585ha$

CHAPTER TEN

1. The expression $px^2 + 12x + 4p + px$

   where p is a constant is a perfect square. **Find** the value of p.

2. If X = 33.5 and y = 33.1 both being correct to one decimal place, **calculate** the maximum possible percentage error in X-y

3. Given that $Sin^2 x + Cos^2 X = 1$ **solve** for X in the equation

   $CosX + Sin X = 1 \ for -180^0 \leq x \leq 180^0$

4.

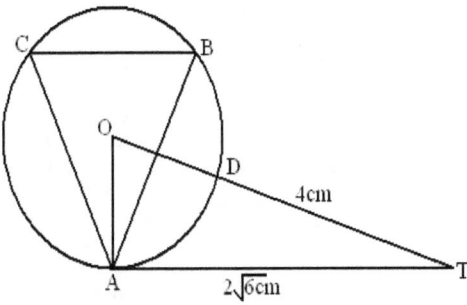

In the figure above, O is the centre of the circle and AT is a tangent to the circle at A. AT = $2\sqrt{6}$ cm and DT = 4cm. **Determine** the value of angle BAT.

5. The cost per head for catering for a party is partly constant and partly varies inversely as the number of people expected. The cost per head for a party of 100 people is Sh.1,860 and that for 180 people is sh. 1,060. **Find** the cost per head for 200 people.

6. The figure below shows a triangular garden in which $\angle ABC = \angle ANB = 90^0$, AN = xm, NC = 4m, $\angle BAN = 30^0$ and $\angle BCN = 60^0$ without using mathematical tables or a calculator, **find** the value of length X leaving your answer in a simplified form.

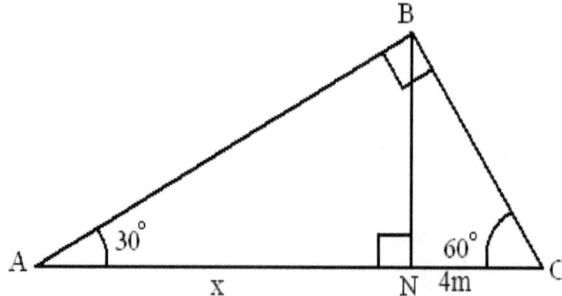

7. (a) **Expand** and **simplify** the expression $\left(4x - \dfrac{y}{2}\right)^5$ up to the third term.

(b) Hence use the expansion in (a) above to **approximate** the value of $(39.6)^5$ correct to 3 significant figures.

8. The graph below shows the linear relation between two variables X and Y connected by the expression $Y = pX^2 + qX$. Using the graph, estimate, to the nearest whole number, the value of

(i) P

(ii) q

9. Use logarithms tables to **evaluate**.

$$\left(\frac{130.9}{27.68 \times 100.9} \right)^{2/3}$$

10. A car valued at Ksh. 600,000 depreciates by 20% in the first year and 10% in the second year. A uniform rate would have depreciated the car in the two years. **Calculate** this uniform rate to 2dp.

11. The difference between the second and fourth terms of an arithmetic sequence is 3.

If the product of the first and the fourth term is 34, **calculate** the value of the first term.

12. **Make** x the subject of the formular:

$$\sqrt{x} = \frac{S - 5a}{6x}$$

13. Five men working six hours a day take eight days to fill a trench. **How** long does it take three men working eight hours a day to complete the same trench?

14. **Draw** using a protractor and ruler only a rectangle ABCD of side 8cm by 3cm. On CD mark two points P_1 and P_2 such that angle APB = angle $AP_2B = 90^0$. Measure P_1P_2.

15. Differentiate $y = 5x - 8x^2 + x^3$. Hence, or otherwise, **determine** the turning points for the curve $y = 5x - 8x^2 + x^3$.

16. The sketch below represents the graph for $y = x^2 - x - 6$. Use the curve and five

trapezia to estimate the area bounded by the

x – axis, y – axis x = 0 and x = 5.

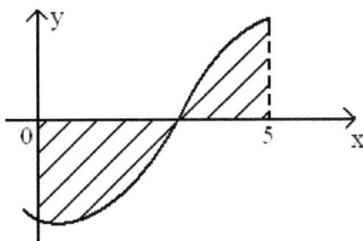

17. Two tanks of equal volume are connected in such a way that one tank can be filled by pipe A in 1hour 20minutes. Pipe B can drain one tank in 3hours 36minutes but pipe C alone can drain both tanks in 9 hours. **Calculate**:

(a) The fraction of one tank that can be filled by pipe A in one hour.

(b) The fraction of one tank that can be drained by both pipes B and C in one hour.

(c) Pipe A closes automatically once both tanks are filled. Assuming that initially both tanks are empty and all pipes opened at once, calculate how long it takes before pipe A closes.

18. An examination involves a written test and a practical test. The probability that a candidate passes the written test is $\frac{6}{11}$ if the candidate passes the written test, then the probability of passing the practical test is $\frac{3}{5}$, otherwise it would be $\frac{2}{7}$

(a) **Illustrate** this information on a tree diagram.

(b) **Determine** the probability that a candidate is awarded
(i) Credit for passing both tests.

(ii)   Pass for passing the written test.

(iii)   Retake for passing one test.

(iv)   Fail for not passing the written test.

19. (a) Conctruct triangle PQR with PQ = 7.2cm, QR = 6.5cm and angle PQR = 48⁰

(b) The locus L1, of points equidistant from P and Q, and locus , L2 of points equidistant from P and R, meet at M. Locate M and measure QM

© A point x moves within triangle PQR such that QX ≥ QM. Shade and label the locus of X.

9. The figure below represents a prism with a cross section of an equilateral triangle of side 8cm and length 12cm, as shown below.

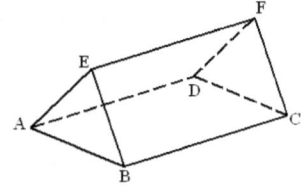

(a) Draw the net of the prism ABCDEF

(b) **Calculate** the angle between the plane ABCD and the line BF.

© M is the midpoint of EF. **Calculate**

(i) The length BM

(ii) The perimeter of triangle BMD.

(d) Calculate the angle between the plane ABM and the base plane ABCD.

21. Give the matrix $A = \begin{pmatrix} -1 & -4 \\ 1 & 3 \end{pmatrix}$

(a)   (i)   **Calculate** A^2 and A^3

(ii) Find the values of the constants p and q for which $A^2 = pA + qI$ where I is the identity matrix.

(iii)The triangle ABC maps onto $A^1B^1C^1$ under the transformation represented by matrix A.

Find the area of triangle ABC if the area of triangle $A^1B^1C^1$ is 21cm²

(b) The figure shows two concetric circles such that the ratio of their radii is 1: 3. If the area of the shaded region is 78.4 square units, **calculate** the area of the larger circle.

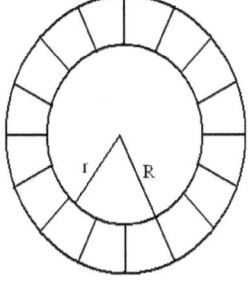

22. A certain uniform supplier is required to supply two types of shirts: one for girls labelled G and the other for boys labelled B. The total number of shirts must not be more than 400. e as to supply more of type G than of type B. However the number of type G shirts must not be more than 300 and the number of type B shirts must not be less than 80. by taking x to be the number of type G shirts and y the number of type B shirts,

(a) Write down in terms of x and y all the inequalities representing the information above.

(b) On the grid provided **draw** the inequalities and shade the unwanted regions.

© Given that type G costs Shs. 500 per shirt and type B costs Shs. 300 per shirt.
   (i) Use the graph in (b) above to **determine** the number of shirts of each type that should be made to maximize profit.

   (ii) **Calculate** the maximum possible profit.

23. (a) The equation of a curve is given by $y = X^3 + X^2 - bx$. **Show** that the value of X at the

      minimum turning point is $\dfrac{-1 + \sqrt{19}}{3}$

(b) The displacement X metres of a particle moving along a straight line after t seconds is

    given by $X = 4t + 2t^2 - t^3$

(i) **Find** its initial acceleration

(ii) **Calculate** the time when the particle was momentarily at rest.

(c) (i) **Find** the values of X where the curve $y = X^2(x - 2)$ crosses the x-axis.

(ii) Hence **find** the area enclosed by the curve $y = X^2(x - 2)$, the lines $x = 0$, $x = 2\frac{2}{3}$ and the

x-   axis.

24. The marks of 50 students in a mathematics test were taken from a form 4 class and recorded
in the table below.

Mark (%)	21-30	31-40	41-50	51-60	61-70	71-80	81-90	91-100
Frequency	2	5	7	9	11	8	5	3

(a) On the grid provided, **draw** a cumulative frequency curve of the data.

    Take: 1cm to represent 5 students on the vertical scale and 1cm to represent 10 marks on

the   horizontal scale.

(b) From your curve in (a) above

(i) **Estimate** the median mark.

(ii) **Determine** the Interquartile deviation.

(iii)**Determine** the 10th to 90th percentile range.

© It is given that students who score over 45 marks pass the test. Use graph in (a) above to **estimate** the percentage of students that pass.

SOLUTIONS TO CHAPTER TEN

NO.	WORKING
1.	$let: PX^2 + 12X + pX + 4p = (KX + q)^2$
	$PX^2 + (12 + P)X + 4P = K^2x^2 + 2kq\,X + q^2$
	$\Rightarrow P = K^2 \ldots\ldots\ldots\ldots(i)$
	$\Rightarrow 12 + p = 2Kq \ldots\ldots\ldots(ii)$
	$\quad 4p = q^2 \ldots\ldots\ldots\ldots(iii) \rightarrow q = \sqrt{4P}$
	$\qquad\qquad\qquad\qquad = \sqrt{4K^2}$
	$\Rightarrow subst\ (i)\ in\ (ii)$
	$\quad 12 + K^2 = 2\,Kq \ldots\ldots\ldots(iv)$
	$\Rightarrow and\ subst.\ (iii)\ in\ (iv)$
	$\quad 12 + K^2 = 2K\sqrt{4K^2}$
	$\qquad K = \pm 2$
	$\therefore P = K^2 = (\pm 2)^2 = 4$
2.	Absol. Error in absol. Error in x + absol. Error in y
	$\qquad x - y = 0.05\ +\ 0.05 = 0.1$
	% error = rel. error X 100
	$= \dfrac{0.1}{33.5 - 33.1} \times 100$
	$= 25\%$
3.	$(Cos\,x + Sin\,x)^2 = (1)^2$
	$Cos^2x + 2Cos\times Sinx + Sin^2x = 1$
	$\therefore 2Cos\,x\ Sinx + 1 = 1$
	$\quad Cosx\ Sinx = 0$
	$\Rightarrow Cos\,x = 0 \qquad or\ Sin\,x = 1$
	$\therefore X = Cos^{-1}(0)\,or\ Sin^{-1}\,(0)$
	$\therefore X = -90^0, -180^0, 0^0, 90^0, 180^0$

4.	$Let\ \ OD = x$
	$(X)^2 + \left(2\sqrt{6}\right)^2 = (x+4)^2$
	$\qquad\qquad X = 1$
	$\tan < AOB = \dfrac{2\sqrt{6}}{1} = 4.8989$
	$\quad < AOB \tilde{\ } 78.5^0$
	$\quad < OAB = \dfrac{[180 - (78.5 \times 2)]}{2}$
	$\qquad\qquad = 23^0$
	$\therefore < BAT = 90 - 23 = 67^0$

5.	$C = k + \dfrac{p}{N}$
	$when: N = 100,\quad C = 1860$
	$\qquad\quad N = 180,\quad C = 1060$
	$1860 = k + \dfrac{p}{100}\ \ \ ....(i)$
	$1060 = k + \dfrac{p}{180}\ ....(ii)$
	$\Rightarrow P = 180,000,\ K = 60$
	$Law:$
	$\qquad C = 60 + \dfrac{180000}{N}$
	$\therefore When\ \ N = 200$
	$\qquad C = 60 + \dfrac{180,000}{200}$
	$\qquad\quad = Sh.\,960$

6.	

	$BN = 4\tan 60^{0} = 4\sqrt{3}\,M$ $\tan 30^{0} = \dfrac{BN}{X}$ $\Rightarrow X = \dfrac{4\sqrt{3}}{1/\sqrt{3}}$ $X = 12m$
7.	(a) $1\,(4x)^{5}\left(\dfrac{-y}{2}\right)^{0} + 5\,(4x)^{4}\left(\dfrac{-y}{2}\right)^{1} + 10\,(4x)^{3}\left(\dfrac{-y}{2}\right)^{2}$ $= 1024x^{5} - 640x^{4}y + 160x^{3}y^{2}$ (b) $\Rightarrow \left(4x - \dfrac{y}{2}\right)^{5} = (39.6)^{5} = (40 - 0.4)^{5}$ $\Rightarrow x = 10,\ y = 0.8$ $on\ subst.;$ $1024(10)^{5} - 640\,(10)^{4}\,(0.8) + 160\,(10)^{3}\,(0.8)^{2}$ $\qquad\qquad = 97382400$ $\qquad\qquad = 97400000 \quad (3sf)$ $\qquad\qquad\qquad or$ $\qquad\qquad\quad (97.4m)$
8.	$Y = pX^{2} + qX$ $\dfrac{Y}{X} = pX + q\ (on\ div.\ by\ X)$ $y = mX + C$ $(i)\ grad,\ m = p = \dfrac{24 - 74}{6 - 1}$ $\qquad\qquad = -10\,(\pm 1)$ $(ii)\ y - \text{int}ercept,\ c = q = 83\,(\pm 1)$

9.	No.	Log
	1309	2.1149
	27.68	1.4422
	100.9	2.0028
		3.4450
		$\overline{2}.6699 \times \frac{2}{3}$

1.298×10^{-1} $\overline{1}.1183$

$$\frac{(-2 + 0.6699) \times 2}{3}$$

$$\frac{-4 + 1.3398}{3}$$

$-6 + 3.3398$

$\overline{1}.1183$

$= 0.1298$

10.

1st year A $= \dfrac{80}{100} \times 600,000 sh$ $=$ Sh. 480,00

2nd year A $= \dfrac{90}{100} \times 480,000$ $=$ Shs. 432,000

A $= P(1 - r/100)^h$

$432,000 = 600,000 (1 - r/100)^2$

$0.72 = (1 - r/100)^2$

$0.8485 = 1 - r/100$

$r/100 - 1 - 0.8485 = 0.1515$

$r = 15.15\%$

11.

$(a + 3d) - (a + d) = 3$

$2d = 3$

$d = \frac{3}{2}$

$a (a + 3d) = 34$

$a^2 + 3a \times \frac{3}{2} - 34 = 0$

$2a^2 + 9a - 68 = 0$

$(2a + 17)(a - 4) = 0$

$a = -8\frac{1}{2}$ or 4

12.	*Either* *or*

12.

Either   *or*

$$x = \frac{(5-5a)^2}{36x^2}$$

$$6x^{3/2} = 5.5a$$

$$x^{3/2} = \frac{5-5a}{6}$$

$$x^3 = \frac{(5-5a)^2}{36}$$

$$x = \left(\frac{5-5a}{6}\right)^{2/3}$$

$$x = 3\sqrt{\frac{(5-5a)^2}{36}}$$

13.

$$8 \ days = 5 \times 6 \ man \ hours$$

$$1 \ day = 5 \times 6 \times 8 \ man \ hours$$

$$Eight \ hour \ day = \frac{5 \times 6 \times 8}{3 \times 8} days$$

$$= 10 \ days$$

14.

$P_1P_2 = 5.1cm \ \pm0.1$

174

15.	$y = 5x - 8x^2 + x^3$ $\dfrac{dy}{dx} = 5 - 16x + 3x^2$ $3x^2 - 16x + 5 = 0$ $(3x^2 - 15x)(-x + 5) = 0$ $(3x - 1)(x - 5) = 0$ $x = \dfrac{1}{3}$ or 5 $x = \dfrac{1}{3}; \; y = 1$ $x = 5; \; y = 50$ *Turning points* $\left(\dfrac{1}{3}, 1\right)$ *and* $(5, 50)$
16.	*Co-ordinates* $(0, -6)(1, -6), (2, -4)(3, 0)(4, 6)$ *and* $(5, 16)$ *neglects signs* $A = \dfrac{1}{2} \times 1\{(6 + 14) + 2(6 + 4 + 0 + 6)\}$ $\qquad = \dfrac{1}{2}\{20 + 2 \times 16\}$ $\qquad = \dfrac{1}{2}\{20 + 32\}$ $\qquad = \dfrac{1}{2} \times 52 = 26 \; sq.units$
NO.	**WORKING**

17.	*Let each* tan*k be x litres* *(a) x is filled in* 1*hr.* 20 min $x = 1\frac{1}{3}\ hrs = \frac{4}{3}$ $\therefore 1hr = \frac{3}{4}x\ filled$

(b) B drains x in 3*hrs* 36 min

$$x = 3\frac{36}{60} = \frac{18}{5}\ hrs$$

$\therefore 1\ hr,\ B\ drains\ \frac{5}{18}x$

  C draws x in $4\frac{1}{2}\ hr$

$x = \frac{9}{2}\ hrs$

1*hr C drains* $\frac{2}{9}x$

Both drains $\dfrac{5}{18}x + \dfrac{2}{9}x$

$$= \frac{5x + 4x}{18}$$

$$= \frac{9x}{18} = \frac{1}{2}x\ of\ a\ \tan k\ filled\ in\ 1hr\ by\ B\ and\ C$$

(c) In 1*hr water in one* tan*k is* $\dfrac{3}{4}x - \dfrac{1}{2}x$

$$= \frac{1}{4}x$$

  so one tan*k fills in* 4*hrs*

  both tan*ks fill in* **8** *hrs*

18.	

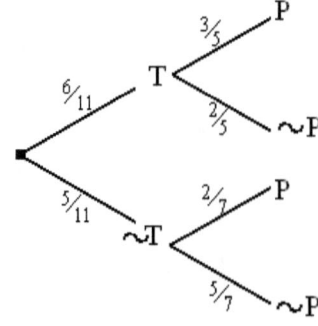

(b) (i) $P(credit) = \dfrac{6}{11} \times \dfrac{3}{5}$

$\qquad = \dfrac{18}{55}$

(b) (ii) $P(pass) = \dfrac{6}{11} \times \dfrac{2}{5}$

$\qquad = \dfrac{12}{55}$

(iii) $P(retake) = \dfrac{6}{11} \times \dfrac{2}{5} + \dfrac{5}{11} \times \dfrac{2}{7}$

$\qquad = \dfrac{18}{55} + \dfrac{10}{77}$

$\qquad = \dfrac{126 + 50}{385}$

$\qquad = \dfrac{176}{385}$

(iv) $P(fail) = \dfrac{5}{11} \times \dfrac{2}{7} + \dfrac{5}{11} \times \dfrac{5}{7}$

$\qquad = \dfrac{10}{77} + \dfrac{25}{75}$

$\qquad = \dfrac{35}{77} = \dfrac{5}{11}$

19.	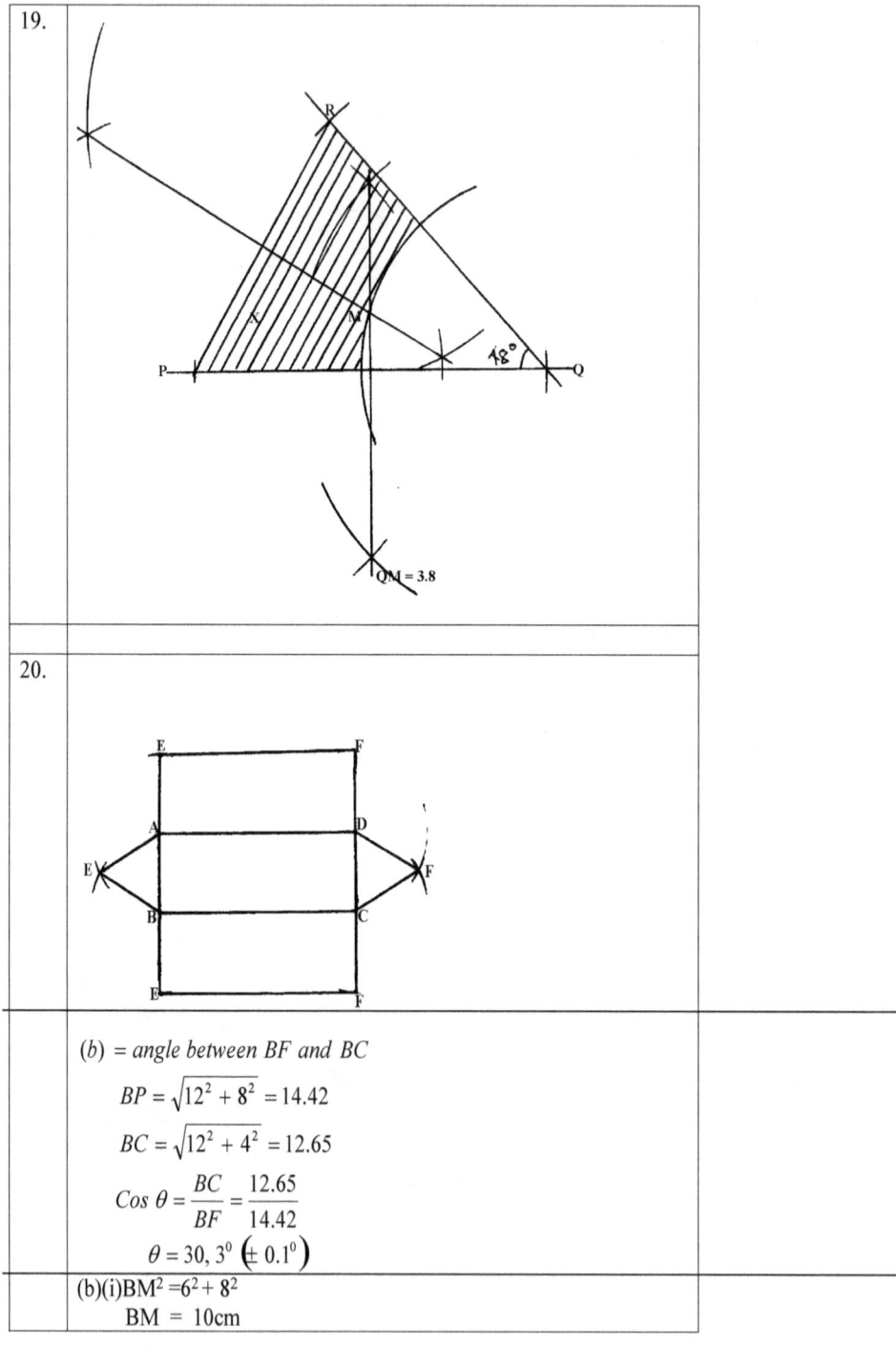	
20.		
	(b) = angle between BF and BC $BP = \sqrt{12^2 + 8^2} = 14.42$ $BC = \sqrt{12^2 + 4^2} = 12.65$ $Cos\ \theta = \dfrac{BC}{BF} = \dfrac{12.65}{14.42}$ $\theta = 30, 3^0\ (\pm 0.1^0)$	
	(b)(i)$BM^2 = 6^2 + 8^2$ $BM = 10cm$	

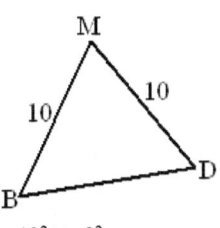

(ii) B

BD2 $= 12^2 + 8^2$

BD $= 14.42$cm

$P = (14.42 + 10 + 10)$ cm

$= 34.42$ cm

(c) = angle between projection of EM on ABCD and the height of the prism

height, h $= \sqrt{8^2 - 4^2}$

h $= \sqrt{48} = 6.928$cm

Projection of EM on ABCD

b = 6cm

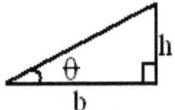

$\tan \theta = \dfrac{h}{b} = \dfrac{6.928}{6} = 1.1546$

$\theta = 49.11^0$

21.	(a) (i) $A^2 = \begin{bmatrix} -1 & -4 \\ 1 & 3 \end{bmatrix}\begin{bmatrix} -1 & -4 \\ 1 & 3 \end{bmatrix}$
	$= \begin{bmatrix} -3 & -8 \\ 2 & 5 \end{bmatrix}$
	$A^3 = A^2 . A = \begin{bmatrix} -3 & -8 \\ 2 & 5 \end{bmatrix}\begin{bmatrix} -1 & -4 \\ 1 & 3 \end{bmatrix}$
	$= \begin{bmatrix} -5 & -12 \\ 3 & 7 \end{bmatrix}$
	(ii) $\begin{bmatrix} -3 & -8 \\ 2 & 5 \end{bmatrix} = p\begin{bmatrix} -1 & -4 \\ 1 & 3 \end{bmatrix} + q\begin{bmatrix} 1 & 0 \\ 0 & 1 \end{bmatrix}$
	$\begin{bmatrix} -3 & -8 \\ 2 & 5 \end{bmatrix} = \begin{bmatrix} -p & -4p \\ p & 3p \end{bmatrix} + \begin{bmatrix} q & 0 \\ 0 & q \end{bmatrix}$
	$p = 2, \quad q = -1$
	(iii) $\Delta\ ABC \Rightarrow \Delta\ A^1\ B^1\ C^1$
	$a..s.f = Det\ A = ^-3 - ^-4 = 1$
	$\therefore \dfrac{Image\ Area}{Object\ Area} = a..s.f$
	$\dfrac{21}{A_0} = 1$
	$A_0 = 21\ cm^2$
	(b) $l.s.f = r : R = 1 : 3$
	$a.. s.\ f = (l.s.f)^2 = 1 : 9$
	ratio of shaded area $= 9 - 1 = 8$
	Area of larger circle $\Rightarrow \dfrac{9}{8} \times 78.4$
	$= 88.2 Sq.\ units$
22.	(a) $x + y \leq 400$
	$x > y$
	$x \leq 300$
	$y \geq 80$
	(b)

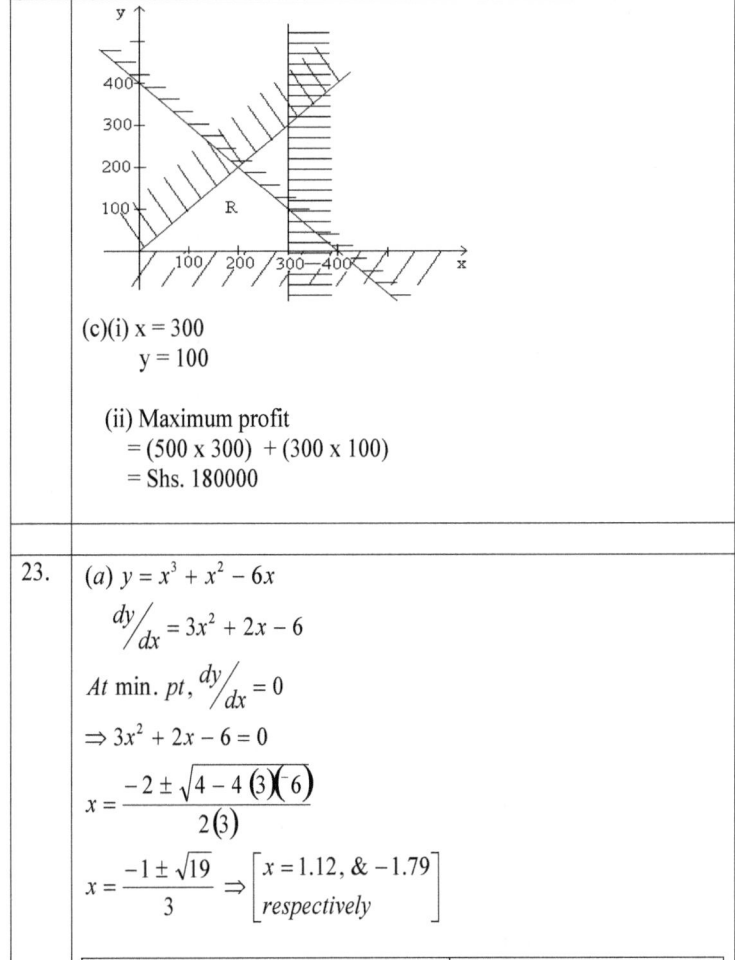

(c)(i) x = 300
   y = 100

(ii) Maximum profit
   = (500 x 300) + (300 x 100)
   = Shs. 180000

---

23. | (a) $y = x^3 + x^2 - 6x$

$$\frac{dy}{dx} = 3x^2 + 2x - 6$$

At min. pt, $\frac{dy}{dx} = 0$

$$\Rightarrow 3x^2 + 2x - 6 = 0$$

$$x = \frac{-2 \pm \sqrt{4 - 4\,(3)(-6)}}{2(3)}$$

$$x = \frac{-1 \pm \sqrt{19}}{3} \Rightarrow \begin{bmatrix} x = 1.12, \,\& -1.79 \\ respectively \end{bmatrix}$$

Nature (x = 1.12)				Nature (x = -1.79		
X	'1	'1.12	'2	-2	-1.79	-1
$\frac{dy}{dx}$	-1	0	+10	+2	0	-1
Sketch	\	—	/	/	—	\
Min. Pt				Max. Pt		

∴ value of x at

$$Min. \; pt \; = \frac{-1 + \sqrt{19}}{3}$$

hence shown

(b) (i) $x = 4t + 2t^2 - t^3$

$V = \dfrac{dx}{dt} = 4 + 4t - 3t^2$, $a = \dfrac{dv}{dt}$

$\qquad\qquad\qquad = 4 - 6t$

Initial acc^n, occurs when $t = 0$

$a = 4 - 4(0\ )$

$\quad = 4\ M/s^2$

(ii) *momentarily at rest when $V = 0$*

$\quad 4 + 4t - 3t^2 = 0$

$\quad (2 - t)(2 + 3t) = 0$

$\quad t = 2\ or\ -1.5$

$\therefore t = 2 Sec$

(c) (i) $X - \text{intercept when } y = 0$

$\quad X^2\ (x - 2) = 0$

$\quad X = 0\ or\ x = 2$

(ii) $A = \displaystyle\int_0^{2\frac{2}{3}} (x^3 - 2x^2)\,dx$

$\qquad = \displaystyle\int_0^2 (x^3 - 2x^2)\,dx + \int_2^{\frac{8}{3}} (x^3 - 2x^2)\,dx$

$\qquad = \left[\dfrac{x^4}{4} - \dfrac{2}{3}x^3\right]_0^2 + \left[\dfrac{x^4}{4} - \dfrac{2x^3}{3}\right]_2^{\frac{8}{3}}$

$\qquad = \left|-\dfrac{4}{3}\right| + \left(\dfrac{4}{3}\right)$

$\qquad = 2\dfrac{2}{3}\ (or\ 2.667)\ sq.\ units$

24. | (a) Completing table & graph

Cumulative Frequency	Upper Limits
2	30.5
7	40.5
14	50.5
23	60.5
34	70.5

42	80.5
47	90.5
50	100.5

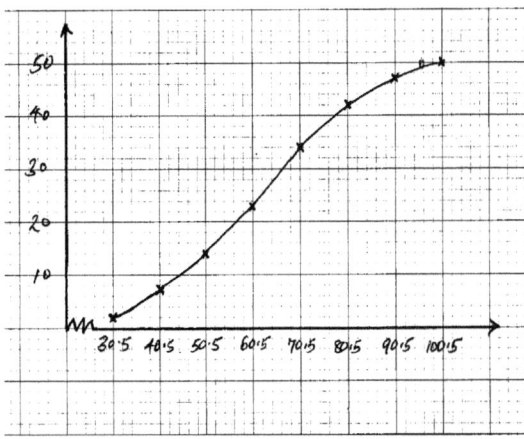

(b) (i) Q2 ≈ 62 (±1) marks
    (ii) Q3 = 75 and Q1 = 48
        Interquartile range
        Q3 - Q1 = 75 - 48 = 27
        ∴Interquartile deviation (Semi – Interquartile range)

$$= \frac{Q3 - Q1}{2}$$

$$= \frac{27}{2}$$

$$= 13.5$$

(iii) P_{90} = 86 marks

P_{10} = 37 marks

∴ The 10 to 90 percentile range

$P_{90} - P_{10} = 85 - 37$

$= 48 \, (\pm 1) \, marks$

(c) From graph, 45 marks is at 21st percentile

i.e 21% of students scored 45 marks and below

∴ % Scoring over 45 marks

$100 - 21 = 79\% \, (\pm 1\%)$

CHAPTER ELEVEN

1. Without using tables or calculators evaluate. (

$$\frac{2\times(-3)+35\div5}{-9+14\div7+4}$$

2. Solve for x and y in $3^{2x} \div 3^{3y} = 2187$ and $2^{3x} \times 2^{6y} = 1$.

3. Three businessmen Njoroge, Mwaura and Kimani contributed a total of sh. 82, 250 to start a business. The ratio of the contribution of Njoroge to Mwaura was **2:3** and that of Mwaura to Kimani was **4:5**. How much did Kimani contribute?

4. Make **t** the subject of the formula.

$$S = 3\sqrt{\frac{3Q-4t}{R+t}}$$

5. Find the equation of the normal to the curve $y = 3x^2 - 9x$ at the point (2, - 6). Express your answer in the form **y = mx + c**

6. In the figure below, if $A\hat{C}B = 50°$, $A\hat{F}B = 35°$ and FC is parallel to AD.

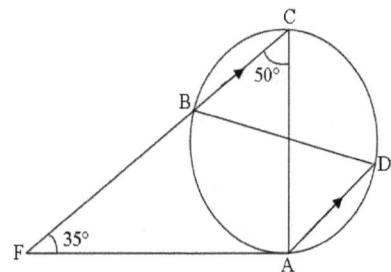

Calculate:

    (i) $A\hat{B}F$

(ii) $A\hat{E}B$

185

7.	Coordinates of points A and B are (2, -3) and (6, 5) respectively. Given that N is a point on AB such that the ratio AN:NB = 1:3, find the coordinates of point N.

8.	Triangle ABC is the image of triangle XYZ when transformed under the matrix $\begin{pmatrix} 3 & -1 \\ 2 & 4 \end{pmatrix}$ .If the area of triangle ABC is 770 cm^2, find the area of triangle XYZ.

9.	In the formula $V = \pi r^2 h$, **r** is increased by 12% and **h** decreased by 15%. Find the percentage change in the value of V.

10.	The fourth term in $(a + x)^3$ is equal to the third term in $(1- 2x)^5$ for the same values of x when expanded in increasing powers of x. Find the value of x.

11.	Graham's money-box contains only sh. 5 coins and sh. 10 coins. There are 24 coins and their total value is sh. 150. Find how many of each kind of coins there are in the box.

12.	Find the distance in nautical miles between two places on the northern hemisphere with positions **N** (60°N, 18°E) and **M** (60°N, 102°W) along the parallel of latitude.

13.	Find the integral values of x which satisfy the compound inequality $3x - 2 \le 8 + x < 12 + 5x$

14.	A committee of three is to be selected from 5 men and 3 women. If the committee is selected at random, find the probability of arriving at two men and one woman in the committee.

15.	Given the line AB below, construct the locus of a point P such that $A\hat{P}B = 60°$ on the upper side of the line segment AB.

A   B

16. From a mathematical experiment it was found out that $\tan 144° \equiv 1 - \sqrt{3}$. Deduce the

value of $\tan 54°$ from this result leaving your answer in the form $\dfrac{a + \sqrt{c}}{b}$ where a, b and c are

integers.

(

17. A matatu left town K at 7.00 a.m and travelled towards town M at an average speed of 60

km/hr. A car left town M at 9.00 a.m and travelled towards K at an average speed of 80 km/hr.

The distance between the two towns is 324 km.

Find;

a) The time each vehicle arrived at their destination.

     (i)   Matatu

     (ii)   Car

b)   (i)   The distance the matatu covered before the car started to move from town M to

town K.

     (ii)   The time the two vehicles met on the way.

c) How far the car was from town K when they met.

18. Mr. Omwega is employed. His basic salary is Kshs. 21, 750 and is entitled to a house

allowance of Kshs 15, 000 and a travelling allowance of Kshs 8, 000 per month. He also claims a

family monthly relief of Kshs 1, 056 per month. Other deductions are;

         Union dues Kshs 200 and

         Co-operative shares Kshs 4, 500 per month.

     The table below shows the tax rates for the year.

Income (Kshs per annum)	Tax rates
1 – 116, 600	10%
116, 161 – 225, 600	15%
225, 601 – 335, 040	20%
335, 041 – 444, 480	25%
Over 444, 480	30%

Calculate;

(a)   Mr. Omwega's annual taxable income.

(b)   The tax paid by Mr. Omwega in the year.

(c)   Mr. Omwega's net income per month.

19.   The frequency distribution table below shows lengths in cm for 41 pieces.

Length	Frequency
5 – 8	4
9 – 12	9
13 – 16	15
17 – 20	8
21 – 24	5

Calculate the:

(a)   Mean.

(b)   Median

(c)   Standard deviation

20.   A particle moves along a straight line such that its displacement **S** metres from a given point is

$S = t^3 - 5t^2 + 3t + 4$ *where t is time in seconds.*

Find:

(a)   The displacement of the particle at t = 5

(b)   The velocity of the particle when t = 5

(c)   The values of t when the particle is momentarily at rest.

(d)   The acceleration of the particle when t = 2.

21.   The field book below gives measurements of a field. The distances are given in metres and

$AF = 100$ m.

	F	
	100	
E 40	80	
	60	D50
C40	40	
	20	B30
	A	

(a)   Using a scale of 1cm represents 10m, draw a map of the field with straight boundary edges.

(b)   (i)   Find the area of the field in square metres.

   (ii)   Determine the area of the field in hectares.

22.   The diagram below (not drawn to scale) shows the cross – section of a hexagonal solid metal prism length 20cm.

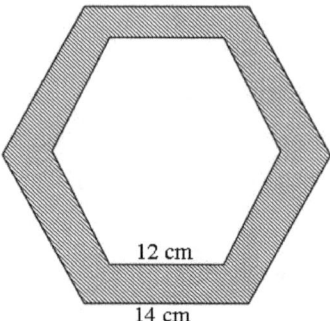

12 cm

14 cm

Calculate;

(a)   the area of the shaded region (Take hexagon to be regular).

(b)   the volume of the material used to make the metal in cm^3.

(c)   If the density of the metal prism is 3.5 g/ cm^3, find its mass in kg.

23.   Two quantities **P** and **n**, are connected by the equation $P = AK^n$, where **A** and **K** are constants. The table below shows some corresponding values of **n** and **P**.

n	2	4	6	8	10
P	9.8	19.4	37.4	74.0	144.4

(a)   State the linear equation connecting **P** and **n**.

(b)   On the grid provided, draw a suitable straight line.

(c)   Use your graph to estimate the value **A** and **k**.

24.   a)   Fill in the table below.

X	0	30	60	90	120	150	180	210	240	270	300	330	360
x + 30	30	60	90	120	150	180	210	240	270	300	330	360	390
sin (x + 30)	0.50	0.87	1.00	0.87	0.50	0.00	- 0.5	- 0.87	- 1.00	- 87	0.50	0.00	0.50
2sin		1.74			1.00			-		-			1.00

190

(x+30)							1.74		1.74		
$1-2\cos 2x$	-1.00	0		3.00		-1.00		2.00		2.00		-1.00

On the same axes sketch the following curves;

$y = 2\sin(x + 30)$ and $y = 1 - 2\cos 2x$.

(c)   Use your graph to solve the equation

$$2\sin(x + 30) + \cos 2x = 1$$

(d)   Find the;

         (i)   Period of $y = 1 - \cos 2x$

         (ii)   Phase angle of $y = 2\sin(x + 30°)$

SOLUTIONS TO CHAPER ELEVEN

1.

$$\left[2 \times (-3)\right] + (35 \div 5)$$

$-9 + (14 \div 7) + 4$

$= \dfrac{-6 + 7}{-9 + 2 + 4}$

$= -\frac{1}{3}$

2.

$3^{2x} \div 3^{3y} = 2187$

$3^{2x} \div 3^{3y} = 3^7$

$3^{2x - 3y} = 3^7$

$2x - 3y = 7$ (i)

$2^{3x} \times 2^{6y} = 1$

$2^{3x + 6y} = 2^0$

$3x + 6y = 0$ (ii)

$2x - 3y = 7$ (i)

$3x + 6y = 0$ (ii)

$4x - 6y = 14$

$\underline{+\ 3y + 6y = 0}$

$7x \qquad = 14$

$x = \dfrac{14}{7}$

$= 2$

$3x + 6y = 0$

$3(2) + 6y = 0$

$6y = -6$

$y = \dfrac{-6}{6}$

$= -1$

$\therefore x = 2$ and $y = -1$

3.

N : M and M : K

2 : 3 $\qquad$ 4 : 5

4(2 : 3) $\quad$ 3 (4 : 5)

8 : 12 $\qquad$ 12 : 15

N : M : K

8 : 12 : 15

Total ratio = 8 + 12 + 15

= 35

	Kimani contributed $\underline{15} \times 82250$	
	35	
	$= \text{sh. } 35{,}250$	
		3
4.	$S^3 = \dfrac{3Q - 4t}{R + t}$	
	$S^3(R + t) = 3Q - 4t$	
	$S^3t + 4t = 3Q - Rs^3$	
	$t(S^3 + 4) = 3Q - RS^3$	
	$t = \dfrac{3Q - RS^3}{S^3 + 4}$	
5.	$y = 3x^2 - 9x$	
	$\dfrac{dy}{dx} = 6x - 9$	
	When $x = 2$	
	$\dfrac{dy}{dx} = 6(2) - 9$	
	$= 3$	
	Gradient f normal at point $(2,-6)$ is $= -1 \div 3$	
	$ = {}^{-1}/_3$	
	$M = {}^{-1}/_3, (2,-6),(x,y)$	
	$\dfrac{y - (-6)}{x - 2} = {}^{-1}/_3$	
	$3y = -x - 16$	
	$y = {}^{-1}/_{3x} - 5\,{}^1/_3$	
6.	(i) $\overset{\wedge}{ABF} = 95°$	
	(ii) $\overset{\wedge}{AEB} = 100°$	
7.		
	$ON = OA + AN$	
	$AN = ¼\,AB$	
	$= ¼\begin{pmatrix} 6 - 2 \\ 5 - (-3) \end{pmatrix}$	
	$= ¼\begin{pmatrix} 4 \\ \end{pmatrix}$	

$$= \begin{pmatrix} 8 \\ 1 \\ 2 \end{pmatrix}$$

ON = QA + AN

$$= \begin{pmatrix} 2 \\ -3 \end{pmatrix} + \begin{pmatrix} 1 \\ 2 \end{pmatrix}$$

$$= \begin{pmatrix} 3 \\ -1 \end{pmatrix}$$

$\therefore$ N is the point (3,-1)

8.	Determinant of $\begin{pmatrix} 3 - 1 \\ 2 - 4 \end{pmatrix}$ is

$= 3 \times 4 - (2x- 1)$

$= 1 2 + 2$

$= 14$

$\dfrac{\text{Area of image}}{\text{Area of object}}$ = determinant

$= \dfrac{770cm^2}{\text{Area of xyz}} = 14$

Area of XYZ $= \dfrac{770cm^2}{14}$

$= 55cm$

9.	New v $=\pi(1.12r)^2 (0.85)h$

% in v $= \left(\dfrac{1.12)^2 (0.85 -1}{1} \right)$ x 100

$\% \Delta$ in v = 6.624 %

10.	Fourth term in $(a + x)^{3} = (1) (x^3) = x^3$

Third term in $(1 - 2x)^5 = 10(1^3) (0-2x)^2 = 40x^2$

Equating $x^3 = 40x^2$

$x = 40$

11.	Let the number of shs.5 coins be x and that of sh. 10 coins y

$x + y = 24$ --- (i)

$5x + 10y = 150$ – (ii)

$5x + 5y = 120$

-

$5x + 10y = 150$

$-5y = -30$

$y = \dfrac{-30}{-5}$

$= 6$

$x + y = 24$

	$x + 6 = 24$ $x \quad = 24 - 6$ $\therefore x = 18$ and $y = 6$
12.	 Longitude difference $= 102 + 18 = 120°$ Distance $= 120 \times 60 \times \cos 60°$ $\qquad\qquad = 3600nm$

13.	$3x - 2 \le 8 + x$ $3x - x \le 8 + R$ $2x \le 10$ $x \le 5$ $8 + x < 12 + 5x$ $8 - 12 < 5x - x$ $\dfrac{-4}{4} < \dfrac{4x}{4}$ $-1 < x$ $\therefore -1 < x \le 5$ Integral values of x are 0,1,2 , 3 4 and 5		
14.	P(2men) $= \,^2/_5$ P(1 women) $= \,^1/_3$ P(2men and 1 woman) $\,^2/_5 \times \,^1/_3$ P(2m and 1w) $= \,^2/_{15}$		

15.

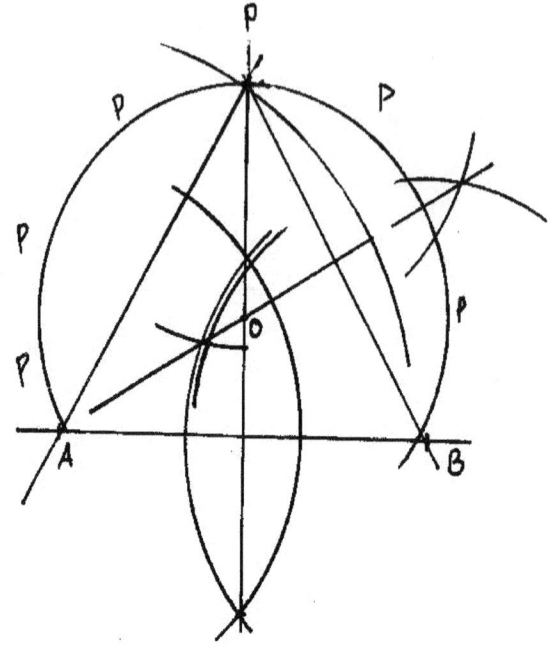

B1 60 constructed ✓ly
B1 2 bisectors and centre of circle identified
B1 Circle arc drawn and marked

3

16. Supplementary ∠ to 144°
$180 - 144° = 36°$
$\tan 36° = -1 + \sqrt{3}$
Complementary angle to 36°
$90 - 36° = 54°$
$\tan 54° = \dfrac{1}{-1 + \sqrt{3}}$

$\dfrac{1}{\sqrt{3} - 1} \quad \dfrac{\sqrt{3} + 1}{\sqrt{3} + 1} \quad = \dfrac{\sqrt{3} + 1}{3 - 1} = \dfrac{\sqrt{3} + 1}{2}$

	$= \dfrac{1 + \sqrt{3}}{2}$
17.	(a) (i) Time = $\dfrac{\text{distance}}{\text{Speed}}$
	$= \dfrac{324\text{km}}{60\text{km/h}}$
	= 5.4 hours
	= 5hrs 24 min
	7.00
	+
	5.24
	12.24 p.m
	(ii) Time = $\dfrac{\text{distance}}{\text{Speed}}$
	$= \dfrac{324 \text{ km}}{80 \text{ km/h}}$
	= 4.05hours
	= 4h 3min
	9.00
	+
	4.03
	13.03 hours or 1.03 p.m
	b)i) Distance = speed x time
	= 60km/h x 2hrs
	= 120km
	ii) Common distance to be
	Covered = 324 – 120
	=204km
	Relative speed = 60 + 80
	= 140km/h
	Time taken to cover common
	Distance = h $\dfrac{204\text{km}}{140\text{km/h}}$ = 1.457hrs = or 1hr 27min
	Time of meeting is
	9.00 +
	1.27
	10.27 a.m
	i) Distance from K is
	= 120km + 60 x 1.457 km
	= 120km + 87.42km
	= 207.42km
18.	a) (21,750 + 15,000 + 8,000) x 12
	= Kshs 537,000

b) Tax slab   Tax pa (Kshs)

1st   116,160 x 10/100   11,616

2nd 109,440 x 15/100   16,416

3rd 109,440 x 20/100   21,888

4th 109,440 x 25/100   27,360

Rem 92,520 x 30 / 100   27,756

Gross tax p.a   105.036

Less family relief pay - 12.672

Net tax p.a   Ksh 92,364.

c) Net income

$$=(21,750 + 15,000 + 8,000) - \frac{(92,364 + 200 + 4,500)}{12}$$

$$= 44,750 - 12,397$$

$$= 32,353$$

19.

Length (cm)	Class – mid(x)	Frequency (f)	fx	fx²	Cumulative frequency
5 – 8	6.5	4	26.0	169	4
9 – 12	10.5	9	94.5	992.25	13
13 -16	14.5	15	217.5	3153.75	28
17-20	18.5	8	148.0	2738	36
21-24	22.5	5	112.5	2531.25	41
		Σf = 41	Σf = 598.5	Σfx² =9584.25	

Mean (x) = $\frac{\Sigma fx}{\Sigma f}$

$$= \frac{598.5}{41}$$

$$=14.5977$$

$$=14.60cm$$

Median position = $(n + 1)/_2$

$$= \frac{41 + 1}{2}$$

$$= 21^{st} \text{ position}$$

Median = 12.5 + 8/15 x 4

$$= 12.5 + 2.133$$

$$=14.633cm$$

$$5 = \frac{\Sigma fx^2}{\Sigma f} - \left(\frac{\Sigma fx^2}{\Sigma f}\right)$$

$$\sqrt{\frac{9584.25}{41} - \left(\frac{598.5}{41}\right)^2}$$

	$= \sqrt{20.66}$ $= 4.545$	

20.	a) $S = 5^3 - 5(5)^2 + 3(5) + 4$
	   $S = 19m$
	b) $V = \dfrac{ds}{dt} = 3t^2 - 10t + 3$
	
	   $= 3(5^2) - 10(5) + 3$
	   $= 28m/s$
	c) Moment at rest $v = 0$
	   $3t^2 - 10t + 3 = 0$
	   $(3t -1)(t -3) = 0$
	   $t = 1/3$ or 3 sec
	
	d) Acceleration when $t = 2$
	   $a = \dfrac{dv}{dt} = 6t - 10$
	
	   $= 2m/s^2$
		1

		10

21.	a) 10m $\longrightarrow$ 1cm
	   100m $\longrightarrow$ $\dfrac{100}{10} = 10cm$
	   40m $\longrightarrow$ $\dfrac{40}{10} = 4\ cm$
	   30m $\longrightarrow$ $\dfrac{30}{10} = 3cm$
	   50m $\longrightarrow$ $\dfrac{50}{10} = 5cm$
		B1
		B1
		B1
		B1

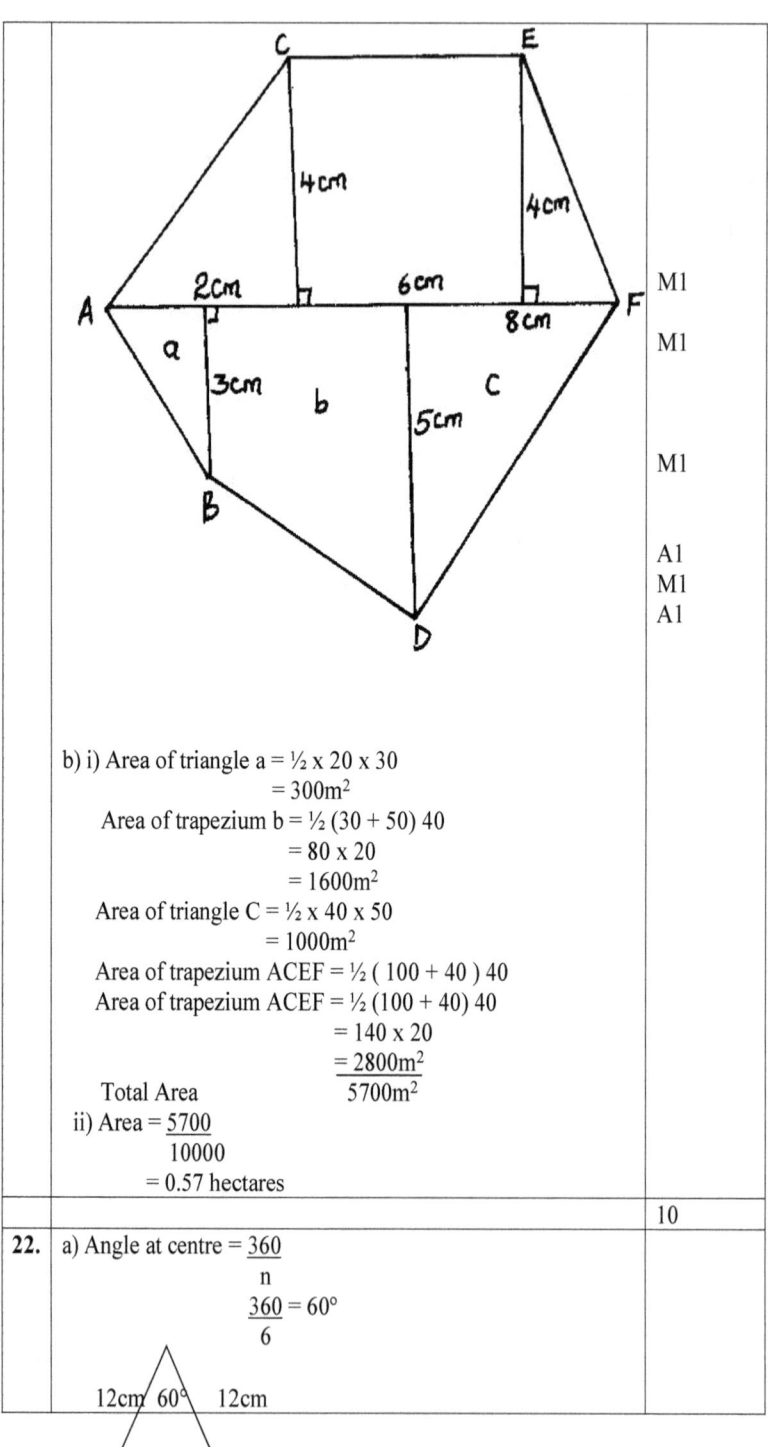

M1

M1

M1

A1
M1
A1

b) i) Area of triangle a = ½ x 20 x 30
                    = 300m²
     Area of trapezium b = ½ (30 + 50) 40
                        = 80 x 20
                        = 1600m²
     Area of triangle C = ½ x 40 x 50
                       = 1000m²
     Area of trapezium ACEF = ½ (100 + 40) 40
     Area of trapezium ACEF = ½ (100 + 40) 40
                           = 140 x 20
                           = 2800m²
     Total Area   5700m²
ii) Area = 5700
           10000
       = 0.57 hectares

| | 10 |

22. a) Angle at centre = 360
                       n
                   360 = 60°
                    6

12cm 60°   12cm

	\ /
	 12cm
	 Shaded area is
	 = 6(½ x 14² sin 60 – ½ x 12² sin 60°)
	 = 6 (98 x 0.866 – 72 x 0.866)
	 = 6(84.868 – 62.352
	 = 69(22.516)
	 = 135.096cm²
	 b) volume = (area of cross – section) x height
	         = 135.096 x 20cm³
	         = 2701.92cm³
	 c) Mass = density x volume
	         = 3.5g/cm³x 2701.92cm³
	         = 9456.72 grams
	         = 9.45672 kg
		M1M1
		 M1
		
		 M1
		 A1
		
		 M1
		 A1
		
		 M1
		 A1
		 B1	

		10

23.

a) $\log p = n \log K + \log A$

b) n

n	2	4	6	8	10
log p	0.99	1.27	1.57	1.87	2.16

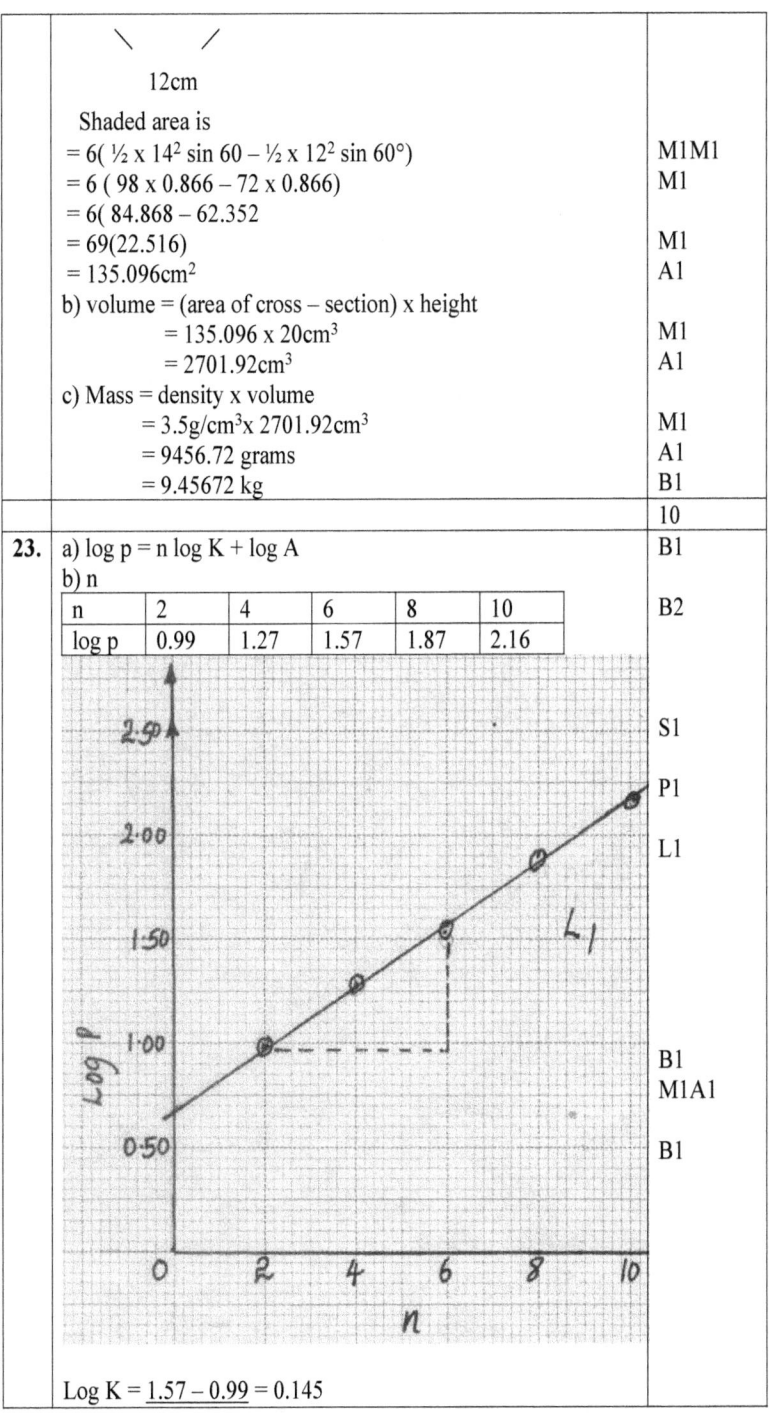

Log K = <u>1.57 – 0.99</u> = 0.145

B1

B2

S1

P1

L1

B1
M1A1

B1

$$\frac{6-2}{}$$

K = 1.40

M1A1
A1

24

X	0	30	60	90	120	150	180	210	240	270	300	330	360
2 sin(x-30)	1.00	1.74	2.00	1.74	1.00	0.00	-1.00	-1.74	-2.00	-1.74	-1.00	0.00	1.00
1-2cos 2x	-1.00	0.00	2.00	3.00	2.00	0.00	-1.00	0.00	2.00	3.00	2.00	0.00	-1.00

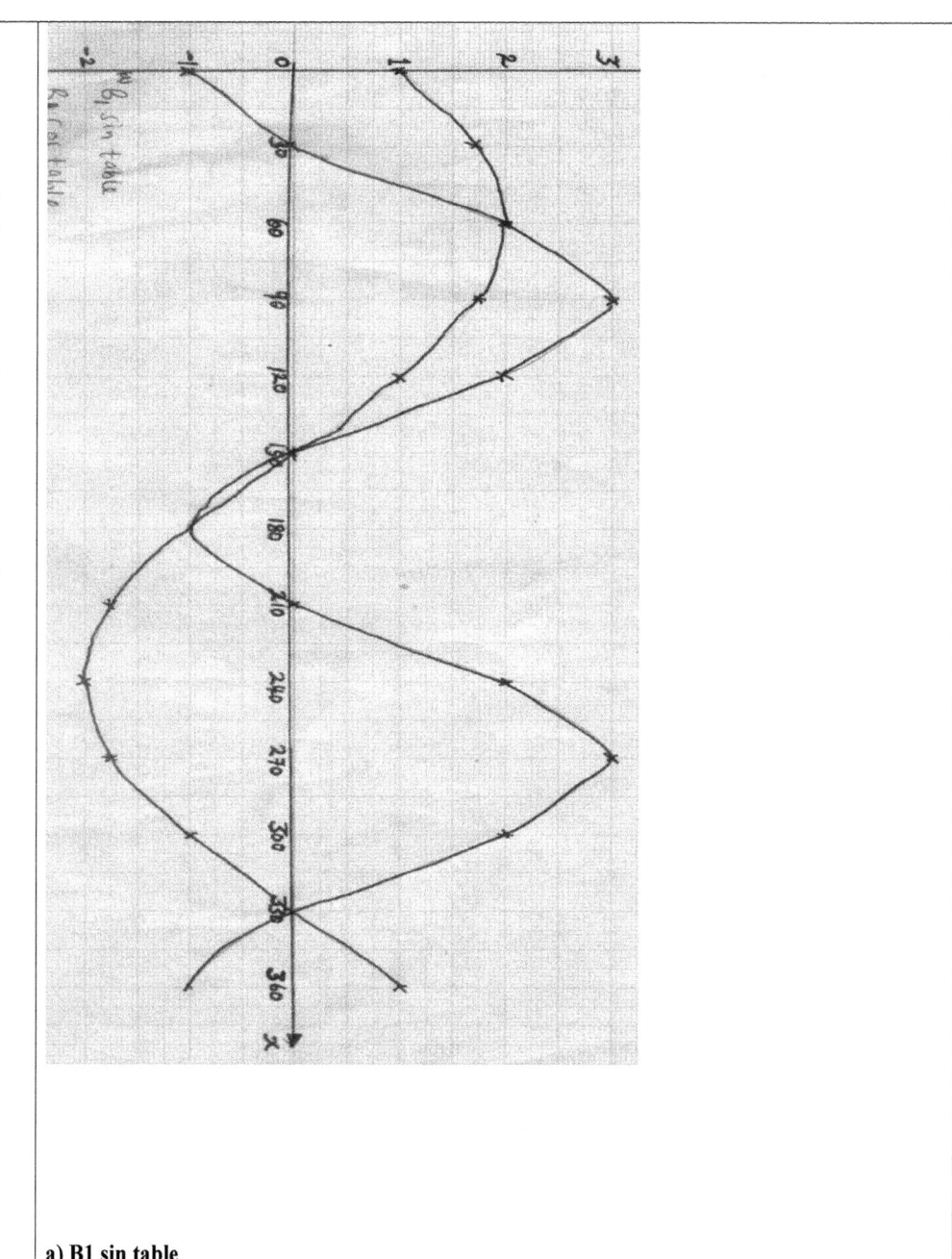

a) B1 sin table
   B1 cos Table
b) S1 Scale
   P1 Plotting

C1 Sin curve

C1 cos Curve

c) $2\sin(x + 30) + \cos 2x = 1$

$60°$ or 150 or 180 or $330°$

d) i) Period of $y = 1 - \cos 2x$ is $180°$ B1

ii) Phase angle of $y = 2\sin(x + 30)$ is 30 B1

CHAPTER TWELVE

1. Simplify the expression below.

$$\frac{2x^2 + 5x - 12}{4x^2 - 9}$$

2. Without using tables, solve the equation.

$$\log_{10}(10x + 5) - \log_{10}(x - 4) = 2$$

3. Given that $\cos \theta = \dfrac{1}{\sqrt{3}}$, find the value of $\dfrac{\tan \theta + \sin \theta}{\cos \theta}$ in its simplest form.

4. Water flows from a pipe at the rate of 250 litres per minute. If the pipe is used to drain a tank full of water measuring 3.2m by 2.5m by 2m, how many minutes would it take to drain the tank completely?

5. In the figure below, triangle $A^I B^I C^I$ is the image of triangle **ABC** under a rotation centre **O**.

By construction, find and label centre **O**. Hence determine the angle of rotation.

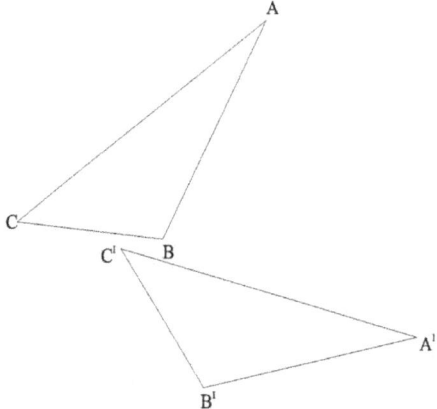

6. The thickness of a mathematics text book is 0.8cm. Eight such books are piled together on a table. Find the limit within which the height of the pile lies.

7. (a) Expand $(1 - 2x)^6$ up to a term including x^3.

   (b) Use the expansion in (a) above to estimate $(0.96)^6$ correct to four decimal places.

8. (a) Find the inverse of the matrix.

$$\begin{pmatrix} 1 & 2 \\ 2 & 2 \end{pmatrix}$$

   (b) Hence solve the simultaneous equations.

   $x + 2y = 21$
   $2x = 34 - 2y$

9. Solve for x and y in the simultaneous equation below.

   $xy + 6 = 0$
   $x - 2y = 7$

10. Points **P**, **Q** and **R** are on the same horizontal ground such that point **P** is due south of **R** and point **Q** is due west of point **R**. **S** is a point at the top of a tower, 20m vertically above **R**. The angles of elevation of **S** from **P** and **Q** are 27° and 35° respectfully. Calculate to the nearest degree, the bearing of **Q** from **P**.

11. Form the three inequalities that satisfy the given region **R**. Given lines **A**, **B** and **C**.

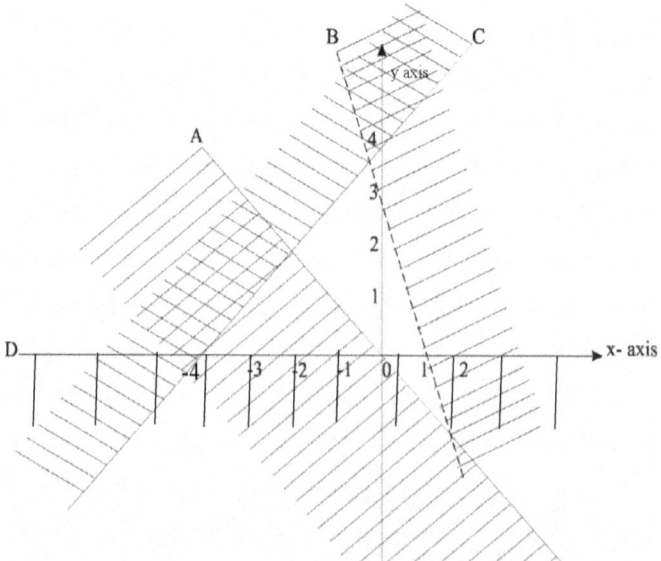

12. In the figure below **PA** is 12mm shorter than **PD**. It is also given that **AB** = 156mm, **CD** = 96mm and **PA** = x mm.

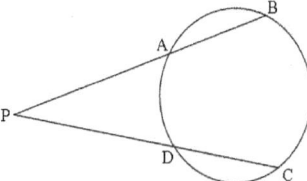

Calculate the value of x.

13. In an arithmetic progression, the 20th term is 92 and the sum of the first 20 terms is 890. Calculate;

   (a)   the first term
   (b)   the common difference

14. The sum of interior angles of a regular polygon is 1440°. Find the number of sides of the polygon.

15. Determine the period and amplitude of the function.

   $y = 4 \sin (2x - 20°)$

16. A body is moving along a straight line and its acceleration after **t** seconds is $(5 – 2t)$ ms^{-2}. Its initial velocity **V**ms^{-1} is 4ms^{-1}. Find **V** in terms of **t**.

17. Bag X contains 2 green marbles and 8 yellow marbles. Bag Y contains 4 green marbles and 5 yellow marbles. A bag is selected at random and two marbles drawn one at a time without replacement.
   (a)   Represent this information on a tree diagram.
   (b)   Find the probability that:

      (i)   They are both green.
      (ii)   They are both yellow and form bag X.
      (iii)   The second ball is yellow.

18.   An aeroplane leaves town **A** (83°N, 155°W) to town **B** (40°N, 25°E) using the shortest route at a speed of 450 knots. (Take $\pi = 3.142$ and radius of the earth **R** = 6370km).

      (a)   (i)   Calculate the distance between **A** and **B** in nautical miles.

            (ii)   Calculate the time taken to travel from town **A** to **B**

(b)   From **B**, the plane flies westwards along the latitude to town **C** (40°N, 13°W). Calculate the distance **BC** in kilometers.

(c)   From town **C**, the plane took off at 3:10 pm towards town **D** (10°N, 13°W), at the same speed. At what time did the plane land at **D**?

19.   (a)   Draw the curve of the function $y = 18 + 3x - x^2$ for $-3 \leq x \leq 6$.

            Use a scale of 2cm to represent 1 unit on x axis and 1cm to represent 1 unit on y axis.

(b)   Find the actual area between the curve and the x-axis.

209

(c)   By using trapezoidal rule with six strips, find the area under the curve between x = -3 and x = 6.

(d)   Find the error introduced by the approximation.

20.   The figure below is a square based pyramid, **ABCDV**, such that **AB**= 7 cm, and **VA**=**VB**= **VC** = **VD**= 9cm.

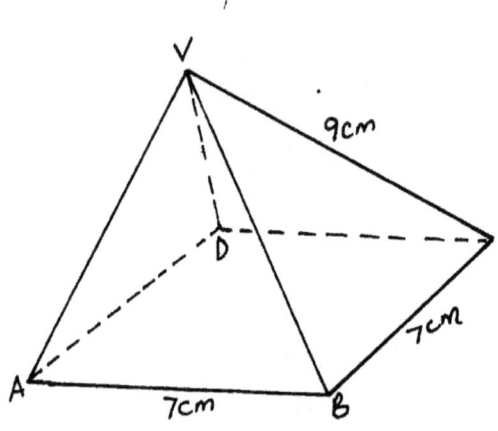

(a)   Find the height of the vertex **V** above the centre of the base.

(b)   Find the angle between **BV** and the base **ABCD**.

(c)   Calculate the angle between the planes **BVC** and **BVA**.

21.   A company receives 470kg of tea leaves in a day. This is packed in 100g, 250g and 500g packets such that the number of 100g packets is twice the number of 250g packets and three times the number of 500g packets.

How many packets of each type are packed in a day.

22. The figure below shows triangle **OAB** in which **OA** is vector a and **OB** is vector b.

Points **D** and **E** are such that $AD = \dfrac{1}{3} AB$ and $OE = \dfrac{1}{3} OA$.

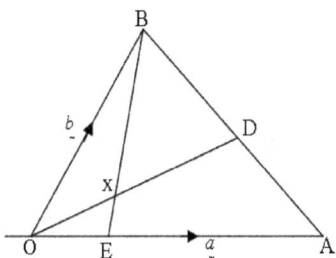

(a)   Express in terms of a and b

     (i) $\overline{OD}$

     (ii) $\overline{BE}$

(b)   If $\overline{OX} = k\overline{OD}$ and $\overline{BX} = h\,\overline{BE}$, where **k** and **h** are constants, express OX in terms of;

     (i)   **k**, a and b

     (ii)   **h**, a and b

(c)   Find the values of **h** and **k**.

23. Using a ruler a pair of compass only;

(a) Draw line **AB** 6cm long. On one side of line **AB**, construct the locus of point **P** such that

$\angle \mathbf{APB} = 45°$

(b) Locate two points $\mathbf{P_1}$ and $\mathbf{P_2}$ on the locus of point **P** such that $\angle \mathbf{AP_1B} = \angle \mathbf{AP_2B} = 45°$ and the area of triangle $\mathbf{AP_1B} = \mathbf{AP_2B} = 12cm^2$.

c) Locate point **Q** on the locus of **P** such that the area of triangle **AQB** is maximum, hence find the maximum area.

24. The points **P** (2, 1), **Q** (4, 1) **R** (4, 3) and **S** (3, 3) are co-ordinates of a quadrilateral.

(a) On the grid provided and using a scale of 2cm to represent 2 units on both axes and taking $-8 \leq x \leq 8$ and $-5 \leq y \leq 8$, draw and label the quadrilateral **PQRS**.

(b)   The matrix $\mathbf{M} = \begin{pmatrix} 1 & 1 \\ 2 & 0 \end{pmatrix}$. Calculate the matrix product $\mathbf{M} \begin{pmatrix} 2 & 4 & 4 & 3 \\ 1 & 1 & 3 & 3 \end{pmatrix}$.

(c)   The image of PQRS under the transformation represented by the matrix, $\mathbf{M}$ is $P^I\, Q^I\, R^I\, S^I$. Draw and label $P^I\, Q^I\, R^I\, S^I$ on the same grid.

(d)   The matrix $\mathbf{N} = \begin{pmatrix} -2 & 1 \\ 0 & 1 \end{pmatrix}$. The image of quadrilateral $P^I\, Q^I\, R^I\, S^I$ under the transformation

represented by the matrix $\mathbf{N}$ is $P^{II}\, Q^{II}\, R^{II}\, S^{II}$. Draw and label $P^{II}\, Q^{II}\, R^{II}\, S^{II}$ on the same grid.

(e)   Determine the matrix that maps **PQRS** directly onto $P^{II}\, Q^{II}\, R^{II}\, S^{II}$.

SOLUTIONS TO CHAPTER TWELVE

1.	$\dfrac{(2x-3)(x+4)}{(2x+3)(2x-3)}$ $= \dfrac{x+4}{2x+3}$
2.	$\text{Log}_{10}\left(\dfrac{10x+5}{x-4}\right) = 2$ $10^2 = \dfrac{10x+5}{x-4}$ $100x - 400 = 10x + 5$ $90x = 450$ $x = 5$
3.	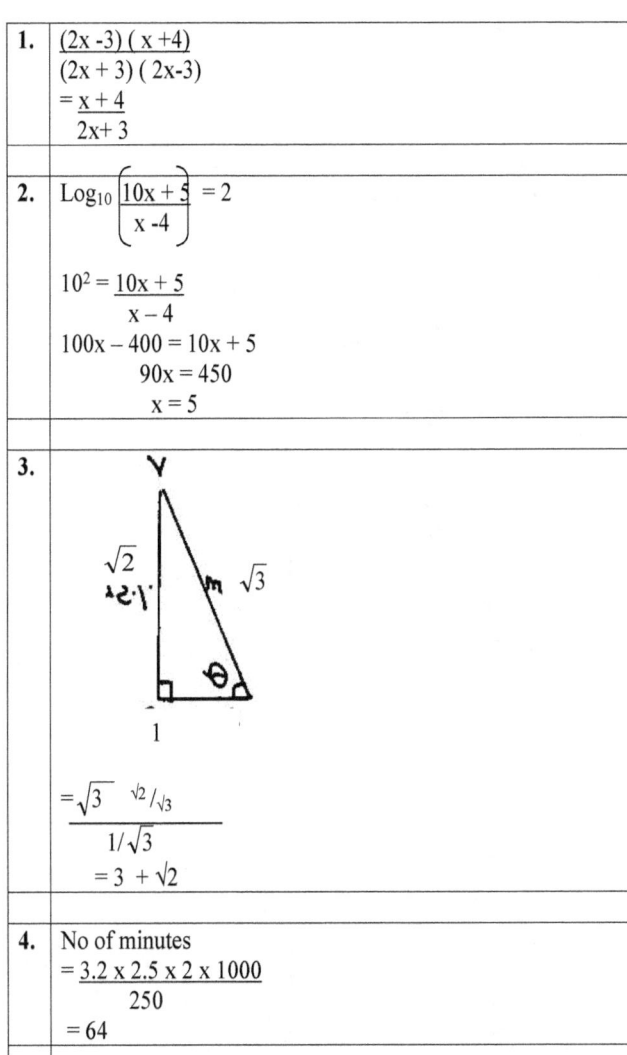 $= \dfrac{\sqrt{3}\quad ^{\sqrt{2}}/_{\sqrt{3}}}{1/\sqrt{3}}$ $= 3 + \sqrt{2}$
4.	No of minutes $= \dfrac{3.2 \times 2.5 \times 2 \times 1000}{250}$ $= 64$

5.

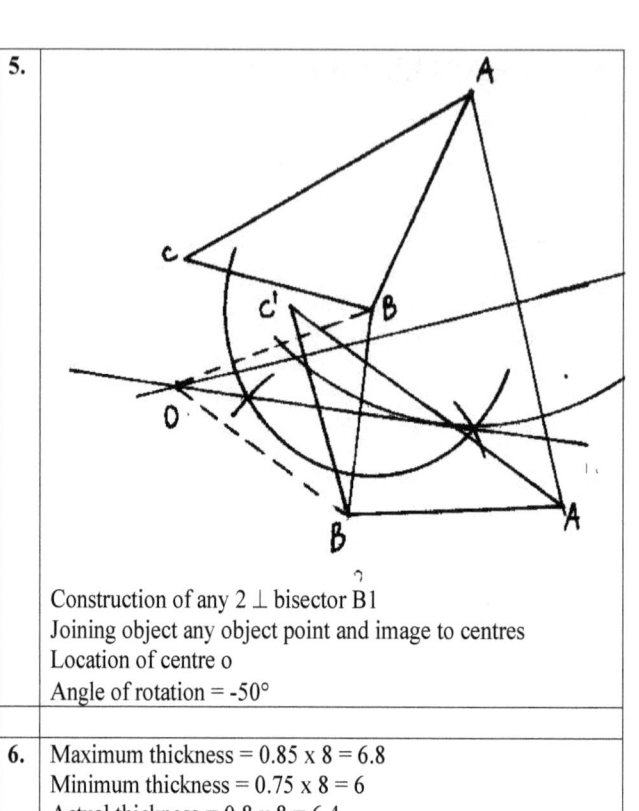

Construction of any 2 ⊥ bisector B1
Joining object any object point and image to centres
Location of centre o
Angle of rotation = -50°

6. Maximum thickness = 0.85 x 8 = 6.8
Minimum thickness = 0.75 x 8 = 6
Actual thickness = 0.8 x 8 = 6.4
A.E = $\underline{6.8\ -6}$ = 0.4
        2
Limit = 6.4 ± 0.4cm

7. a) $1-2x + 4x^2 - 8x^3$
    1 6   15   20
$1 - 12x + 60x^2 - 160x^3$
b)$(0.96)^6 = (1- 0.04)^6$
   $-2x = -0.04$
     $x = 0.02$
    $1 - 12(0.02) + 60(0.02)^2 - 160(0.02)^3$
    $1 - 0.24 + 0.024 - 0.00128$
    $= 0.77272$
     0.7727

8. a) Det (1 x 2) – (2 x 2) = -2
    Inverse = $^1/_{-2}\begin{pmatrix}2 - 2\\ 2\ \ 1\end{pmatrix}$ = $\begin{matrix}-1\ \ 1\\ 1\ -\frac{1}{2}\end{matrix}$)
b)$\begin{pmatrix}1\ 1\\ 1-\frac{1}{2}\end{pmatrix}$ $\begin{pmatrix}1\ 2\\ 2\ 2\end{pmatrix}$ $\begin{pmatrix}x\\ y\end{pmatrix}$ = $\begin{pmatrix}-1\ 1\\ 1-\frac{1}{2}\end{pmatrix}$ $\begin{matrix}21\\ 34\ x\ 21 + -\frac{1}{2}\ x\ 34\end{matrix}$

$$\begin{bmatrix} 1 & 0 \\ 0 & 1 \end{bmatrix} \begin{matrix} x \\ y \end{matrix} = \begin{matrix} -1 \times 21 + 1 \times 34 \\ 1 \times 21 + -\frac{1}{2} \times 34 \end{matrix}$$

$x = 13$

$y = 4$

9.	$x = 7 + 2y$

$x = 7 + 2y$

$(7 + 2y)\, y + 6 = 0$

$7y + 2y^2 + 6 = 0$

$2y^2 + 7y + 6 = 0$

$2y^2 + 4y + 3y + 60 = 0$

$2y(y + 2) + 3(y + 2) = 0$

$(2y + 3)(y + 2) = 0$

When $y = -2$, $x = 3$

$y = {}^{-3}/_{21}$, $x = 4$

10.

$QR = \dfrac{20}{\tan 35°}$

$= 28.56m$

$RP = \dfrac{20}{\tan 27°}$

$= 39.25m$

$\theta = \tan^{-1} \left(\dfrac{RP}{QR} \right)$

$= 53.96m$

11. $L_A\ y + x \geq O$

$L_B\ y + 3x < 3$

$L_c\ y - x \geq 4$

	L_D $y \geq O$
12.	PB x PA = PC x PD $(x + 156)x = (x + 108)(x + 12)$ x = 36mm
13.	a) $92 = a + 19d$ $\quad 89 = 2a + 19d$ $\quad\quad 3 = -a$ $\quad\quad a = -3$
	b) $a + 19d = 92$ $\quad -3 + 19d = 92$ $\quad\quad\quad d = 5$ $\quad$ or $a + 19d = 92$ $2a + 19d = 89$ $\quad\quad\quad d = 5$
14.	$90°(2n - 4) = 400°$ $\quad 2n - 4 = \dfrac{1440°}{90°}$ $\quad 2n - 4 = 16$ $\quad \dfrac{2n}{2} = \dfrac{20}{2}$ $\quad n = 10$ Number of sides = 10
15.	Let $(x - 10) = \theta$ then $y = 4 \sin 2\theta$ Period $= \dfrac{360°}{2} = 180°$ Amplitude = 4
16.	$V = \int (5 - 2t)\,dt$ $V = 5t - t^2 + c$ Substitute, v = 4 when t = O $4 = 5 \times 0 - 0^2 + c$ $\quad\quad C = 4$ $V = 5t - t^2 + 4$
17.	a)

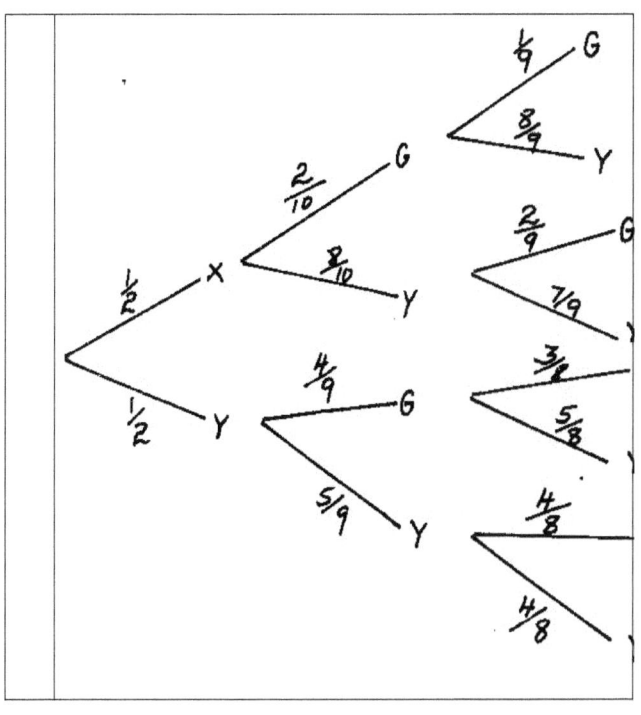

b)i) $P_{(XGG)}$ or $P_{(YGG)}$
$= \frac{1}{2} \times \frac{2}{10} \times \frac{1}{9} + \frac{1}{2} \times \frac{4}{9} \times \frac{3}{8}$
$= \frac{308}{3420}$ (or 0.090)

ii) $P_{(xyy)} = \frac{1}{2} \times \frac{8}{10} \times \frac{7}{9}$
$= \frac{14}{45}$ (or 0.311

iii)$P_{(XGY)}$ or $P_{(XYY)}$ or $P_{(YGY)}$ or $P_{(YYY)}$
$= \frac{1}{2} \times \frac{2}{10} \times \frac{8}{9} + \frac{1}{2} \times \frac{8}{10} \times \frac{7}{10} + \frac{1}{2} \times \frac{4}{9} \times \frac{5}{8} + \frac{1}{2} \times \frac{5}{9} \times \frac{4}{8}$
$= \frac{16}{180} + \frac{56}{200} + \frac{28}{180} + \frac{20}{144}$
$= \frac{597}{900}$ (or 0.663)

18.

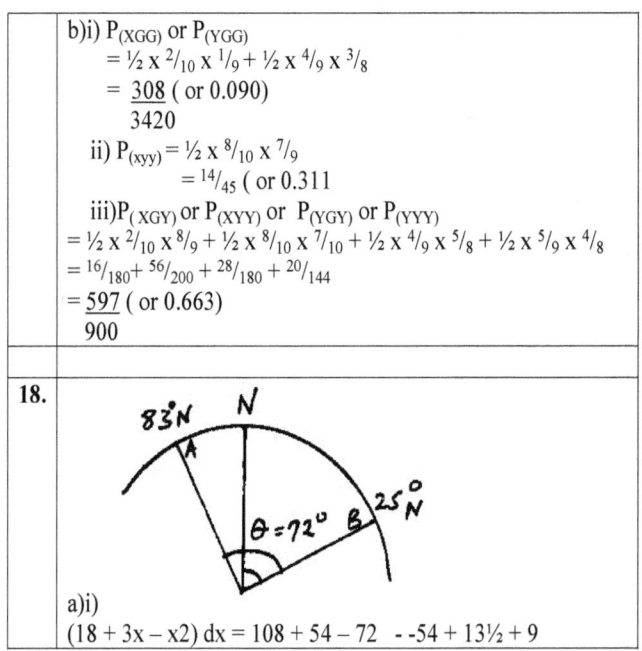

a)i)
$(18 + 3x - x2) dx = 108 + 54 - 72 \quad - -54 + 13\frac{1}{2} + 9$

AB = $\dfrac{72 \times 3.142 \times 2 \times 2 \times 6370}{360 \times 1.853}$

    = 4,320nm

ii) time = $\dfrac{4,320}{450}$

      = 9hrs 36minutes

b) BC = $\dfrac{38}{360} \times 6,370 \times \text{Cos}40° \times 2 \times 3.142$

    = 3,236.76 km

c) CD = 60 x 30 = 1800nm

    time taken = $^{1800}/_{450}$ = 4 hours

    Arrival time = 3:10 pm + 4hours

           = 7 : 10pm

---

19. a)

x	-3	-2	-1	0	1	2	3	4	5	6
y	0	8	14	18	20	20	18	14	8	0

b)A $\int_{-3}^{6} \left(18 + 3x - x^2\right) dx = \left[108 + 54 - 72\right] -$

$\left[-54 + 13^{1}/_{2} + 9\right]$

   121.5

c) Area = ½ (a+ b)h

   ½ (0 + 11.4) 1.5 = 8.55

   ½ (11.4 + 18) 1.5 = 22.05

½ (18 + 20) 1.5 = 28.50

½ (18 + 20) 1.5 =28.50

½ (18 + 11) 1.5 = 21.0

½ (11 + 0)1.5 = $\underline{8.25}$

      Total 116.85

---

d)Error = Actual – Approximate		
= 121.5 – 116.85		
= 4.65		
	10	

19.

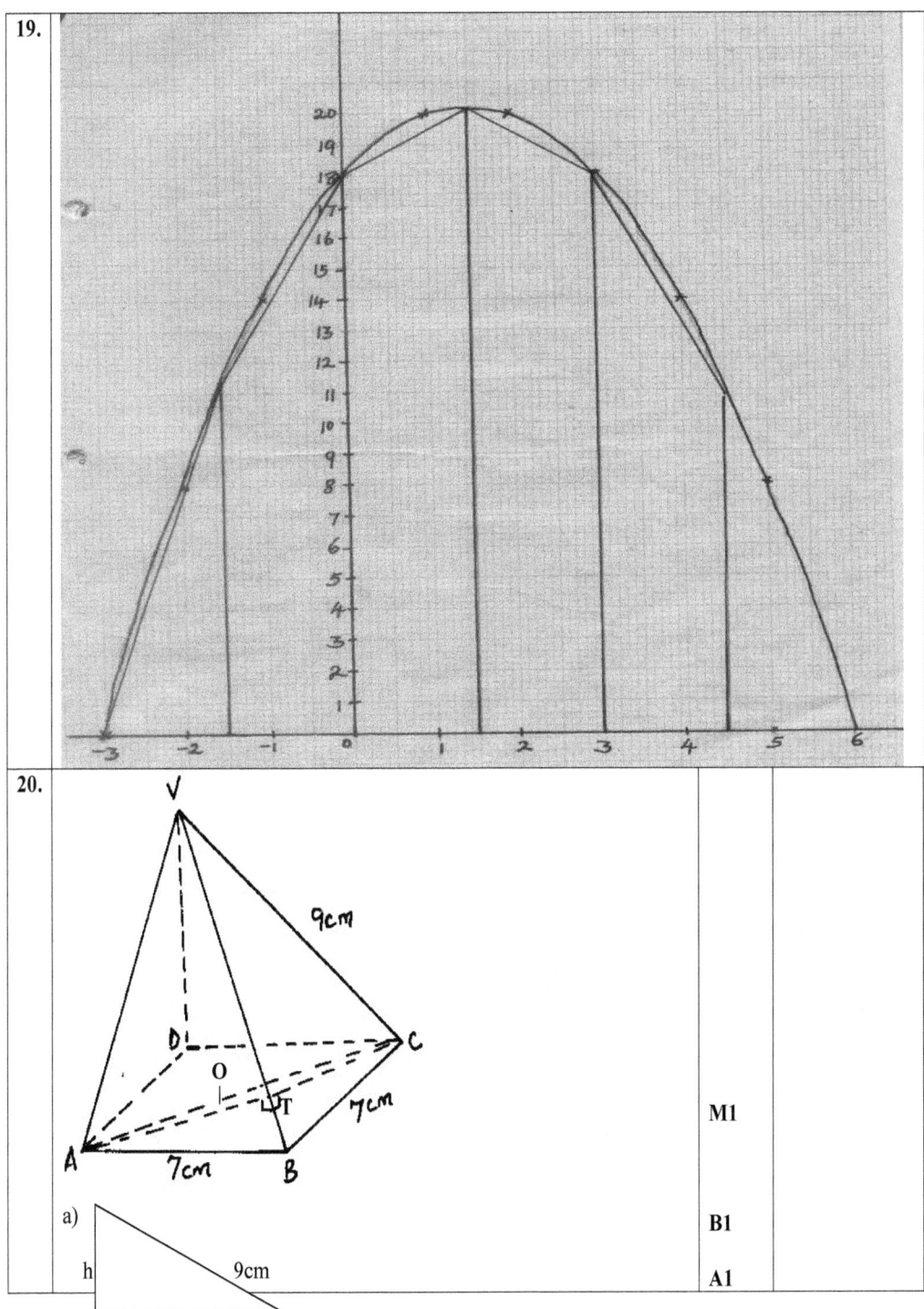

20.

a)

M1

B1

A1

221

O   C

$h = (VC)^2 - (OC)^2$ **M1**

             A1

$= \sqrt{81 - (OC)^2}$

$OC = \sqrt{\dfrac{7^2 + 7^2}{2}}$

     $= 4.95\text{cm}$

             A1

$\therefore h = \sqrt{81 - 24.5025}$

    $= 7.52\text{cm}$

b)

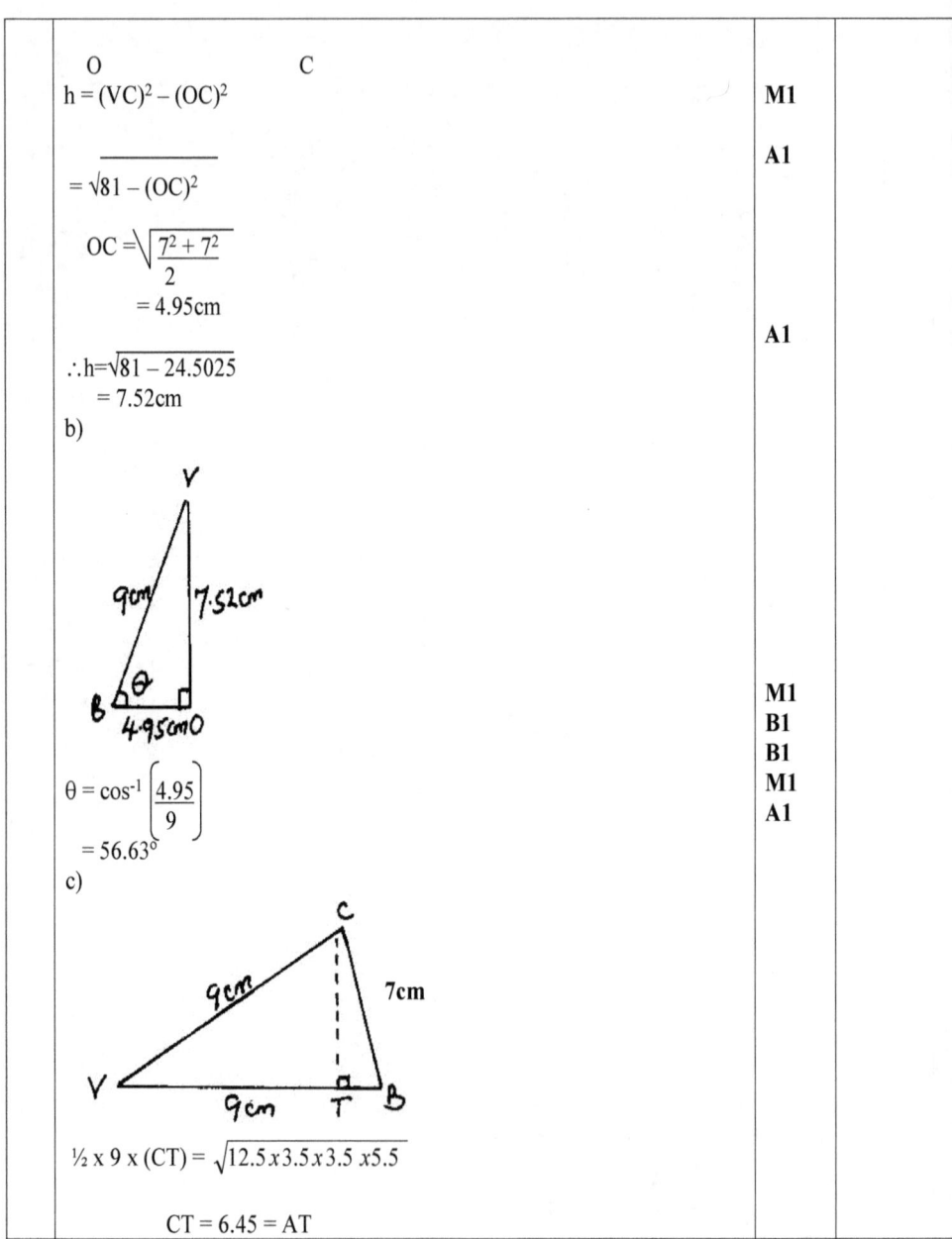

$\theta = \cos^{-1}\left[\dfrac{4.95}{9}\right]$ **M1**
                             B1
                             B1
   $= 56.63°$ **M1**
                             A1

c)

$\tfrac{1}{2} \times 9 \times (CT) = \sqrt{12.5 \times 3.5 \times 3.5 \times 5.5}$

       $CT = 6.45 = AT$

$(9.90)^2 = 4.45^{22} + 4.45^2 - 2 \times 4.45 \times 4.745 \cos\theta$

or

$\theta \sin^{-1}\left(\dfrac{4.95}{6.45}\right)$

$= 100.2°$

			10	
21.	Let the number of 250g packets be x 100g 250g 500g 2x x $^1/_3$ of 2x $2x \times 100 + x \times 250 + ^2/_3 X \times 500 = 470 \times 1000$ $600x + 750x + 1000x = 1410000$ $\quad\quad 2350x = 1410000$ $\quad\quad\quad\quad x = 600$ No of packets for $\quad 100g = 1200$ $\quad 250 = 600$ $\quad 500g = 400$			**B1B1** **M1A1** **M1** **M1** **A1** **B1** **B1** **B1**	
			10	

22.	a) i) OD = OA + 1/3 AB
	$\quad\quad\quad = a + ^1/_3 (b - a)$
	$\quad\quad\quad = ^2/_3a + ^1/_3b$
	  ii) BE = $^1/_3a - b$
	b) i) OX = k (2/3 a + 1/3kb
	$\quad\quad\quad = ^2/_3ka + ½ kb$
	$\quad\quad\quad\quad ☐\quad\quad\quad ☐$
	  ii) OX = OB + BX
	$\quad\quad\quad\quad b + hBE$
	$\quad\quad\quad = b + h(1/3a - b)$
	$\quad\quad\quad = 1/3h a + (1-h)b$
	c) $^2/_3ka + ^1/_3kb = ^1/_3ha + (1-h)b$
	$\quad\quad ☐\quad\quad\quad ☐\quad\quad\quad ☐$
	$\quad ^2/_3 ka = ^1/_3a \Rightarrow ^2/_3k = ^1/_3h$
	$\quad\quad ☐\quad\quad\quad ☐\quad\quad\quad ☐$
	~
				B1
				B1
				B1
				M1
				B1
				M1
				M1	

$k = \frac{1}{2} h$			
$\frac{1}{3}kb = (1-h)b$	**M1**		
$\frac{1}{3}k = (1-h)$ Substitute h for k			
$\frac{1}{3}(\frac{1}{2}h) = 1 - h$			
~			
$\frac{1}{3}(\frac{1}{2}h) = 1 - h$	**B1**		
$\frac{1}{6}h + h = 1$	**B1**		
$h = \frac{6}{7}$			
$k = \frac{3}{7}$			
	10		

23.

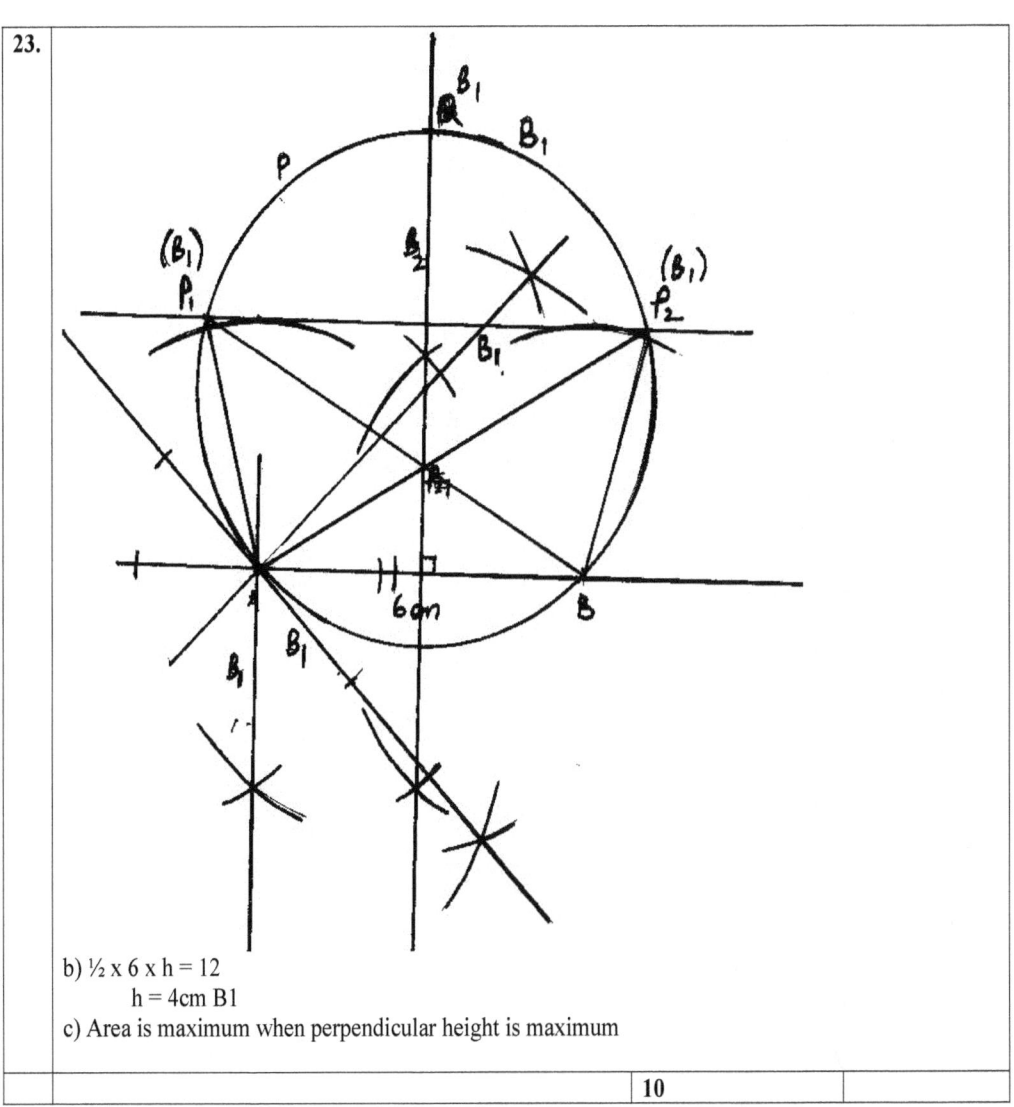

b) ½ x 6 x h = 12

     h = 4cm B1

c) Area is maximum when perpendicular height is maximum

10

24. $\begin{pmatrix} 1 & 1 \\ 2 & 0 \end{pmatrix} \begin{pmatrix} 2 & 4 & 4 & 3 \\ 1 & 1 & 3 & 3 \end{pmatrix}$	$\begin{pmatrix} 3 & 5 & 7 & 6 \\ 4 & 8 & 8 & 6 \end{pmatrix}$	B1B1	
$\begin{pmatrix} -2 & 1 \\ 0 & 1 \end{pmatrix} \begin{pmatrix} 3 & 5 & 7 & 6 \\ 4 & 8 & 8 & 6 \end{pmatrix}$	$\begin{pmatrix} -2 & -2 & 6 & 6 \\ 4 & 8 & 8 & 6 \end{pmatrix}$	B1B1	
$\begin{matrix} a & b \\ c & d \end{matrix} \begin{pmatrix} 2 & 4 & 4 & 3 \\ 1 & 1 & 3 & 3 \end{pmatrix}$	$\begin{pmatrix} -2 & -2 & 6 & -6 \\ 4 & 8 & 8 & 6 \end{pmatrix}$	B1	
$2a + b = -2$ (for ✓ equation containing a and b formation of equals) $-2 = 0$ $a = 0$ $\qquad b = -2$ $2c + d = 4$ $4c + d = 8$ $2c + d = 4$ $4c + d = 8$ $-2c = -4$ $c = 2$ $d = 0$ single matrix $\begin{pmatrix} 0 & -2 \\ 2 & 0 \end{pmatrix}$ for writing the ✓ matrix		B1	

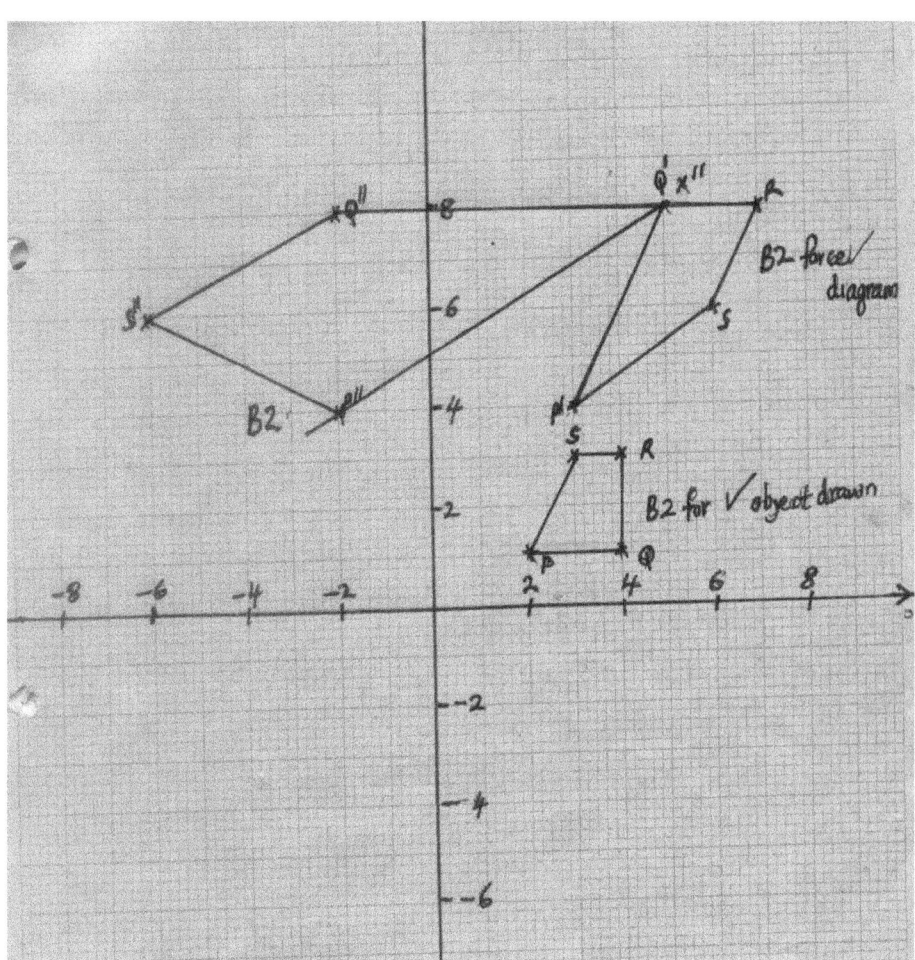

B2- force/ diagram

B2 for ✓ object drawn

227

CHAPTER THIRTEEN

1. Express the number 1470 and 7056 each as a product of its prime factor. Hence evaluate

$\dfrac{1470^2}{\sqrt{7056}}$, leaving the answer in prime factor form.

2. The figure below shows a solid made by pasting two equal regular tetrahedron

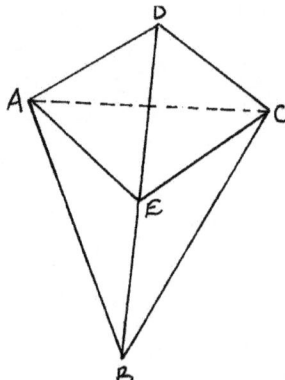

    (a)   Draw a net of the solid

    (b)   If each face is an equilateral triangle of sides 5cm, find the surface area of the
solid

    3. Solve the following equation for x

$$9^x + 3^{2x-1} - 1 = 107$$

4. In the given figure, GJ is parallel to HICD FH is parallel to CJ. Angle AGB = 30^0 and angle AHC = 63^0. Find angle GCJ.

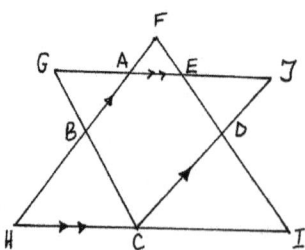

5. A trouser whose marked price is Kshs 800 is sold to a customer after allowing him a discount of 13%. If the trader makes a profit of 20%, find how much the trader paid for the trouser.

6. If $\sin \alpha = 5t$ and $\cos \alpha = 6t$, find t

7. Make x the subject of the formula, $P = 2\pi\sqrt{\dfrac{1 + x^n}{w}}$

8. Simplify the expression
$$\frac{4x^2 - y^2}{3y^2 - 7xy + 2x^2}$$

9. Use mathematical tables to evaluate to 4 decimal places.
$$\frac{1}{\sqrt{0.24680}} + 0.14682^3$$

10. Find by calculation the sum of all the interior angles in the figure ABCDEFGHI below

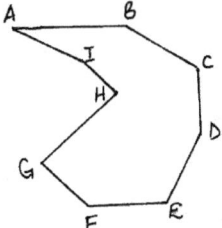

11. Triangle ABC has vertices A at (1, 4) and B at (0, 1) and the line y = x + 1 and its axis of symmetry. Find the co-ordinates of C.

12. The position vectors of points x, y and z are, $x = t\,\tilde{i} + 2\,\tilde{j} : y = 2\,\tilde{i} + 3\,\tilde{j}$, and $2 = 3\,\tilde{i} + 4\,\tilde{j}$ respectively. If the points x, y and z are collinear, find the value of t

13. A Kenyan bank buys and sells foreign currencies at the exchange rates shown below

	Buying	Selling
	Kshs	Kshs
1 Euro	147.56	148.00
1 US Dollar	74.22	74.50

An American arrived in Kenya with 20,000 Euros. He converted all the Euros to Kenya shillings at the bank. He spent Ksh 2510200 while in Kenya and converted the remaining Kenya shillings into US dollars at the bank. Find the amount in Dollars that he received.

14. The figure drawn below is a square ABCD of sides 36.75cm. The shaded area is formed out of two segments. DCB and DKB. Find the area of the shaded region.

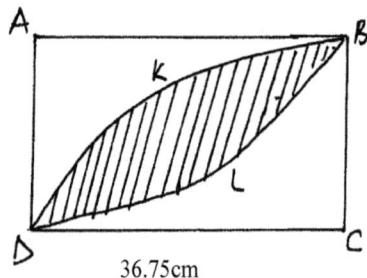

36.75cm

15. Solve the following simultaneous equations, give the exact values of a and b.

$\log (a - b + 1) = 0$ and $\log (ab) + 1 = 0$

16. A tangent from an external point A(7, 5) meets a circle centre O whose equation is $x^2 + 4x + y^2 - 5 = 0$ at point T, find the length of the tangent AT

17. Patients who attended a clinic in one week were grouped by age as in the table

Age years	$0 \leq x < 5$	$5 \leq x < 15$	$15 \leq x < 25$	$25 \leq x < 45$	$45 \leq x < 75$
No. of patients	14	41	60	70	15

   (a) Estimate the mean age
   (b) Draw a histogram to represent the distribution.

18. In the year 2001, the price of a sofa set in a shop was Kshs. 12000,
   (a) Calculate the amount of money received from the sale of 240 sofa sets that year
   (b) (i) In the year 2002, the price of each sofa set increased by 25% while the number of sets sold decreased by 10%. Calculate the percentage increase in the amount received from the sales
       (ii) If at the end of year 2002 the price of each sofa set changed in the ratio 16:15.
           Calculate the price of each sofa set in the year 2003
   (c) The number of sofa sets sold in the year 2003 was P% less than the number of sofa sets sold in the year 2001. Calculate the value of P given that the amount received from sale in the two years were equal

19. A bus left Mombasa and traveled towards Nairobi at an average speed of 60Km/h. after 2 ½ hours a car left Mombasa and traveled along the same road at an average speed of 100km/h. If the
   distance between Mombasa and Nairobi is 500km
   Determine;
   (a) (i) The distance of the bus from Nairobi when the car took off
       (ii) The distance the car traveled to catch up with the bus

(b)   Immediately the car caught up with the bus the car stopped for 25 minutes. Find the new average speed at which the car traveled in order to reach Nairobi at the same time as the bus.

20.   A hemispherical bowl has internal and external radii 30cm and 32cm respectively.
    (a)   Calculate the volume of the material used in making the bowl.
    (b)   The bowl is made of copper of density 9g/cm³. Calculate the mass of the copper used to   make the bowl.
    (c)   Find the total cost of making the bowl if copper cost shs 450 per kg

21.   Four towns P, Q, R and S are such that Q is 168km from P on a bearing of 063^0. R is 288km on a bearing of S 30^0E from Q. S is due west of R on a bearing of 161^0 from P using the scale of 1cm to represent 40km,
    (a)   Show the relative position of P, Q, R and S.

    (b)   From the diagram, find
        (i) The bearing of S from Q
        (ii) The bearing of P from R
    (iii) The distance of PS and SR

22.   A transformation T_1 maps $\triangle ABC$ whose vertices are A(-2, 0), B(1, -2) and C (0, 1) onto $\triangle A^1B^1C^1$ whose vertices are A^1(2, 4), B^1(4, 1) and C^1(1, 2). Another transformation T_2 maps the same $\triangle ABC$ onto $\triangle A^{11}B^{11}C^{11}$ whose co-ordinates are A^{11}(4,2), B^{11}(1, 4) and C^{11}(2, 1). Another transformation T_3 maps $\triangle ABC$ onto $\triangle A^{111}B^{111}C^{111}$ such that A^{111}(-4,0), B^{111}(2,-4) and C^{111}(0,2).
    (a)   On the same axis draw the triangles ABC, $A^1B^1C^1$ and $A^{11}B^{11}C^{11}$

23. Two circles of radius 7cm each with centres x and y. The circles touch each other at point Q. Given that $\angle AXD = \angle BYC = 120^0$ and lines AB, XQY and DC are parallel. Calculate the area of

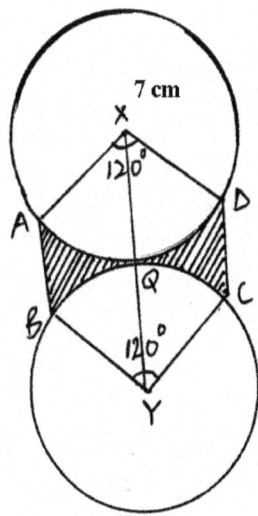

(a) Minor sector $\angle AQD$ (Take $\pi = {}^{22}/_7$)

(b) The trapezium $\angle ABY$

(c) The shaded regions

24. Bag x contains 672 sweets, bag Y 504 sweets and bag Z 360 sweets, all of different types. The sweets are to be shared out among a group of students in such a way that each student gets the same number of each type.

(a) (i) Find the largest possible number of students.

    (ii) What is the total number of sweets received by each student

(b) A number A leaves a reminder of 5 when divided by 29 and also a remainder of 5 when divide by 23. Find the least value of A.

SOLUTIONS TO CHAPTER THIRTEEN

1.	$1470 = 2 \times 3 \times 5 \times 7^2$ $7056 = 2^4 \times 3^2 \times 7^2$ $$\frac{\left(2 \times 3 \times 5 \times 7^2\right)^3}{\sqrt{2^4 \times 3^2 \times 7^2}}$$ $3 \times 5^2 \times 7^3$
2.	a) 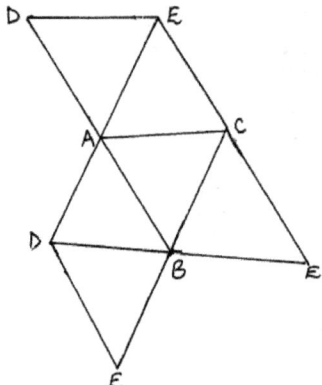 b) Area of a $= 6 \times \frac{1}{2} \times 5 \times 5 \sin 60$ $\qquad = 64.95 \text{cm}^2$
3.	$9x + 32x - 1 - 1 = 107$ $9x + \dfrac{9^x}{3} = 108$ $9x \left(\dfrac{4}{3}\right) = 108$ $9^x = \dfrac{3 \times \overset{27}{\cancel{108}}}{\cancel{4}} = 9^2$ $\quad x = 2$
4	$\angle AGB = 30°$; $BCH = 30°$ (alt $\angle$' s) $\angle BHC = 63$, $\angle DCI = 63$ But $\angle BCD = 117° - 30° - 63° = 87°$ $\therefore \angle GCS = BCD = 87°$

5.	$S.P = \dfrac{87}{100} x800 = Shs696$ net cost price be shs x $x = \dfrac{696 \; x \, 100}{120}$ $= Shs \; 580$
6.	$\sin \alpha = 5t, \cos \alpha = 6t$ $= Tan \, \alpha = \dfrac{\sin \alpha}{\cos \alpha} = \dfrac{5t}{6t} = 0.8333$ $\alpha = Tan\text{-}1(0.8333) = 39.80$ $Sin \; 39.80 = 5t \text{ or } 39.8 = 6t$ $t = 0.128$
7	$p = 2\pi \sqrt{\dfrac{1+x^n}{w}}$ $\left(\dfrac{p}{2\pi}\right)^2 = \dfrac{1+x^n}{w}$ $\dfrac{p^2}{4\pi^2} = \dfrac{1+x^n}{w}$ $\dfrac{wp^2}{4\pi^2} - 1 = x^n$ $x^n = \dfrac{wp^2 - 4\pi^2}{4\pi^2}$ $x = \dfrac{\sqrt[n]{wp^2 - 4\pi^2}}{4\pi^2}$
8	$\dfrac{4x^2 - y^2}{3y^2 - 7xy + 2x^2}$ $\dfrac{(2x + y)(2x - y)}{(y - 2x)(3y - x)}$ $-\dfrac{(2x+y)}{3y - x}$

9.	$\dfrac{1}{\sqrt{0.2468}} + 0.14682^3$
	$\sqrt{0.2468} = \left(24.68 \ x \ 10^{-2}\right)^{1/2}$
	$= 4.9679 \ x \ 10^{-1}$
	Reciprocal $= 10 \ \ x \ 0.2012$
	$0.14682^3 = 1.4682^3 \ x \ \ 10^{-3}$
	$\qquad\qquad = 3.162 \ x \ 10^{-3}$
	$\qquad\qquad$ sum $= 2.012 + \ 0.003162$
	$\qquad\qquad = 2.015162$
10	Sum of int $\angle$'s $= (\ 2n \ - 4) \ 90$
	$n = 9$
	$(\ 2 \ x \ 9 - 4) \ 90$
	$\ 1260$
11.	
	Mid pt AC $= \frac{1}{2} \begin{pmatrix} x+1 \\ y+4 \end{pmatrix} = \begin{pmatrix} \dfrac{x+1}{2} & \dfrac{y+1}{2} \end{pmatrix}$
	Gradient of AC $= -1$ perpendicular lines
	hence $\dfrac{y-4}{x-1} = -1 \ \ y = 5 - x$
	at contact mid pt
	$5 - x = x + 1$
	$4 = 2x \ = 1 \ x = 2 \ \ y = 3$
	but at mid pt co ordinates
	$\left(\dfrac{x+1}{2} \quad \dfrac{y+4}{2} \right) = \left(2,3\right)$
	$x = 3 \ - d \ y = 2$
	pt C (3,2)
12.	$x = ti \ + 2j$
	$y = 2i \ + 3j$
	$z = 3i \ + 4 \ j$

	$xy = \begin{pmatrix} 2-t \\ 3-2 \end{pmatrix} = \begin{pmatrix} 2-t \\ 1 \end{pmatrix}$ $y2 = \begin{pmatrix} 3-2 \\ 4-3 \end{pmatrix} = \begin{pmatrix} 1 \\ 1 \end{pmatrix}$ Collinear pts $k\begin{pmatrix} 2-t \\ 1 \end{pmatrix} = \begin{pmatrix} 1 \\ 1 \end{pmatrix}$ $k(2-t) = 1 \ or \ k = 1$ $t = 1$
13.	Amount of Kshs = 2000 x 147.50 $\qquad$ = 2,951,200 Balance 2,951,200 – 2,510,200 $\qquad$ = Ksh 441000 US Dollars = $\dfrac{441000}{74.5}$ $\qquad$ = 5919.463087
14.	Shaded area $\left(\dfrac{90}{360} \times 3.142 \ x36.75^2 - \frac{1}{2} x36.75^2 \right)^2$ $\left(1060.866844 - 675.28125\right)^2$ $771.171188 \ or 771.17$
15.	$Log(a - b + 1) = 0$ & $\log(ab) + 1 = 0$ $Log(a - b +) = \log 1$ & $\log(ab) + \log 10 = \log 1$ $a - b + 1 = 1$ & $10 \ ab = 1$ $a - b = 0$ $a = b \qquad$ & $10a^2 = 1$ $a = \sqrt{\frac{1}{10} = b}$

$$a = \frac{\sqrt{10}}{10} = b$$

| 16 | $x^2 + 4x + y + y^2 = 5 + 4 = 9$ |

$(x + 2)^2 + y^2 = 3^2$

Centre (-2,0)   radius 3

$\sqrt{9^2 + 5^2} = \sqrt{81 + 25} = \sqrt{105}$

$(AT^2 = (ac)2 - r^2$

$105 - 3 = 102$

$AT = 10.1$

17.	Age	Fre(f)	x	fx	cf	midpts
	0 - 5	14	2.5	35	14	
	5 – 15	41	10	410	55	fx✓
	15 - 25	60	20	1200	115	
	25 - 45	70	35	2450	185	Σ fx✓
	45 - 70	15	57.5	862.5	200	
		200		4957.5		

a) $x = \dfrac{4957.5}{200}$

$= 34.78759\ 24.79)$

b)
Age	f	fd
0 - 5	14	$^{14}/_5 = 28$
4 - 15	41	$^{41}/_{10} = 4.1$
15 - 25	60	$^{60}/_{10} = 6.0$
25 - 45	70	$^{70}/_{20} = 3.5$
45 - 70	75	$^{15}/_{25} = 0.6$

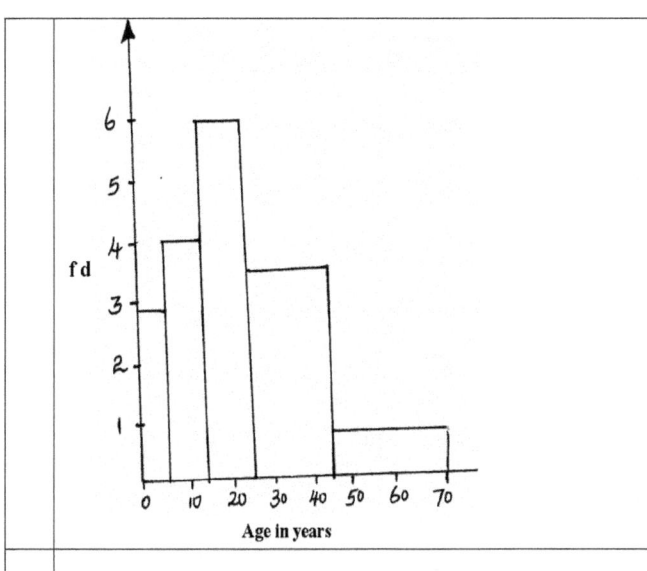

Age in years

| 18. | a) Number of sofa sets sold = 240 at 12000
amount of sales = 240 x 12000
= 2880000
b) (2) New price per sofa set
12000 x $\frac{125}{100}$ = shs 15000
Number sold = 240 x $\frac{90}{100}$ = 216
Amount received = 216 x 15000
= 3240 000
Increase in the amount is 3240 000 − 2880000
= Sh 360000
& increase $\frac{360\ 000}{2880000}$ x 100
= 12.5%
ii) New price is 2003
$\frac{15\ 000}{15}$ x 16   sh 16000 |

Amount of sofa sets is
240 − $\frac{240p}{100}$
Amount received is
16000(240 − 2.4p) = 2880000
- 2.4p = 180 - 240
= 62.4
p % = 62.4%

19.	a) i) The distance the bus traveled when the car took off

19.

a) i) The distance the bus traveled when the car took off

$2\frac{1}{2}$ h x 60

= 150km

ii) Let x be the distance traveled by the car to catch up time

taken by the car to catch up $= \dfrac{x}{100} h$

bus has traveled $\dfrac{x-150}{60}$

$\dfrac{x}{100} = \dfrac{x-150}{60}$

$100x - 15000 = 60x$

$x = 375$ km

b) Distance remaining after car catches up is $500 - 375 = 125$ km

Time taken by bus $\dfrac{125}{60}$

$\left(\dfrac{125}{60} h - 25\min \right)$ *to cover the same distance*

cars average speed

$\dfrac{125km}{\dfrac{125}{60}h - 25\min} = \dfrac{125km}{\dfrac{125}{60} - \dfrac{25}{60}}$

$\dfrac{125 \times 60}{100} = 75 km.h$

20.

a) $Vol = \frac{1}{2}\left(\dfrac{4}{3}\pi r^3 - \dfrac{4}{3}\pi r^3 \right)$

$\dfrac{2}{3}\pi \left(32^3 - 30^3 \right)$

$= \dfrac{2}{3} \dfrac{22}{7} 5768$

$\dfrac{253.792}{21}$

$= 12085.3 cm^3$ or $12085 \frac{1}{3} cm^3$

b) mass of copper used without handles

$= 12085 \dfrac{1}{3} \times 9g$

= with hadles included

= 12085 $\dfrac{1}{3 \, x \, 9 \ x \ 125}$

= 135.96kg
c) Cost 13596 x 450
   = shs   61182

21.

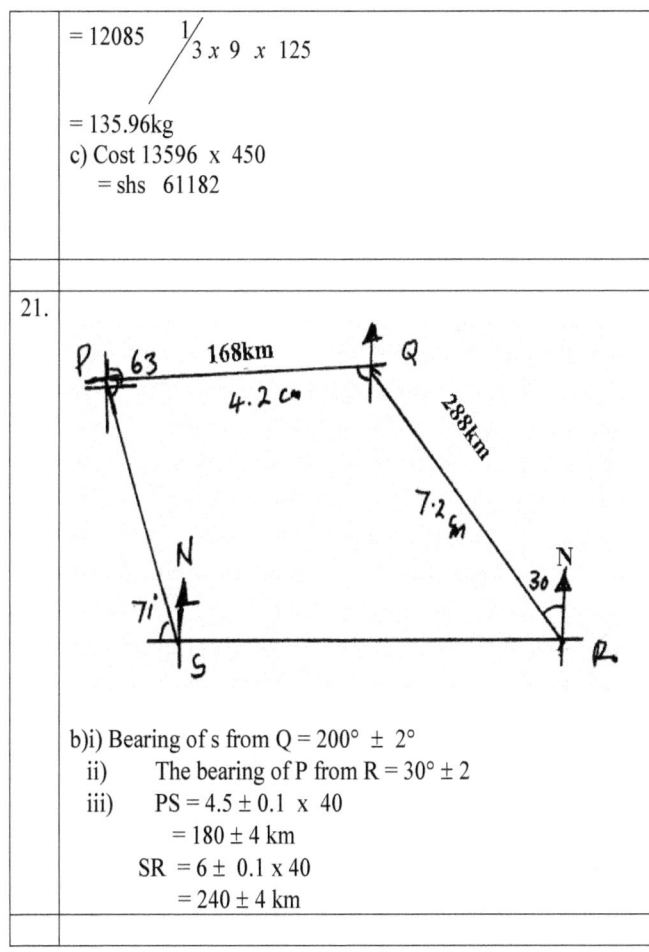

b)i) Bearing of s from Q = 200° ± 2°
   ii)   The bearing of P from R = 30° ± 2
   iii)   PS = 4.5 ± 0.1 x 40
            = 180 ± 4 km
         SR = 6 ± 0.1 x 40
            = 240 ± 4 km

22.	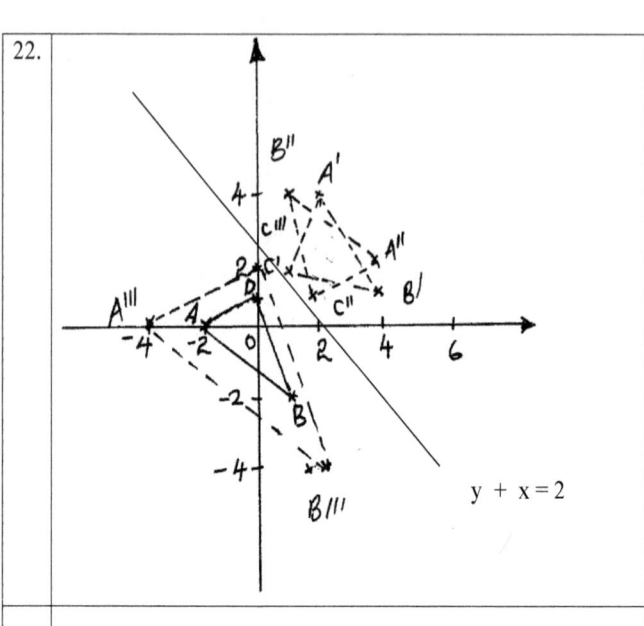

$$y + x = 2$$

i) T_2 Reflection in the line $y + x = 2$

ii) T_2 is a half line about (1,1)

iii) T_3 is an enlargement centre (0,0) and s.f 2

---

23. a) Area of minor sector XAQD

$$\frac{120}{360} \, x \, \frac{22}{7} \, x \, 7^2$$

$$= 1/3 \, x \, 22 \, x \, 7$$

$$= 51 \tfrac{1}{3} \, cm^2$$

b) Drop a perpendicular from A and B to meet x y at M C & N respectively.

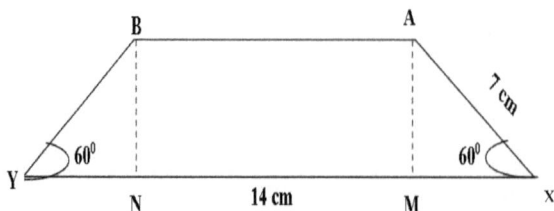

Using AMX; Am = 7 sin 60
$$= 6.062$$
and XM $\quad = 7 \cos 60$
$$= 3.5$$
AB = 14 – 2 x 3.5 = 7
Area of A x yb = ½ x 6.062(14 + 7)
$$= 3.031 \text{ x } 21$$
$$= 63.65 \text{cm}$$
c) Area of ABYCDX is 63.65 x 2 = 127.3
   Area of minor sectors AXD and BYC
   511/3 x 2
   $= 102\frac{2}{3}$
Shaded area = 127.3 - 102.67
$$= 24.63$$

---

24. a)i)

2	672	504	360
2	336	252	180
2	168	126	90
3	84	63	45
	28	21	15

G.C.D = 2^3 x 3 = 24

ii) Sweets obtained by a student from bag x $\underline{672}$ = 28
$$24$$

   Sweets obtained by a student from bag y = $\underline{504}$ = 21
$$24$$

   Sweets obtained by a student from bag z = $\underline{360}$ = 15
$$24$$

Total number of sweets received by each student
$$= 28 + 21 + 15 = 64$$

CHAPTER FOURTEEN

1. Use logarithms to evaluate

    $$\frac{5.246x\log 0.2349}{0.06364^{1/2}}$$

2. Make x the subject of the formula

    $$M = \sqrt{\frac{x-p}{p(1+px)}}$$

3. In the figure below, O is the centre of the circle. Chord AB and CD intersect at X. CX = 8 cm,

    XD = 5 cm and XB = 4cm

    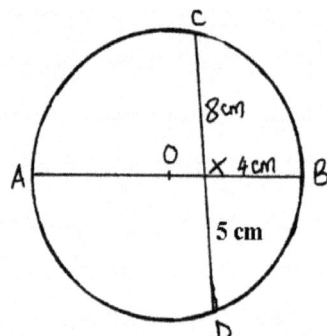

    Calculate the length of AX, hence find the radius of the circle.

4. Simplify completely the expression

    $$\frac{6x^2y^2 - 4xy^3}{9x^3y - 4xy^3}$$

5. Given A, B and C are matrices as shown below

    $$A = \begin{pmatrix} 2 & 3 \\ {}^-5 & 0 \end{pmatrix} \quad B = \begin{pmatrix} 3 & {}^-1 \\ 4 & 2 \end{pmatrix} \quad C = \begin{pmatrix} 1 & {}^-4 \\ 5 & 1 \end{pmatrix}$$

    Determine matrix D given by D = AC − B and hence its inverse

6.   In calculating the volume of a cone a student had an error of 0.2 % in π, 1.2 % in the height and 0.4 % in the radius, calculate the percentage error involved in calculating the volume.

7.   A line L1 passes through points A(3,3) and B $\left(5, ^{-}1\right)$ and it is bisected by line L2. Determine the equation of line L2

8.   Determine the quartile deviation of the data
      18,15,21,19,17,22,21

9.   a)   Complete the table below by filling the blank spaces

x	0	0.5	1	1.5	2
$y = 40 - 3x^2$	40		37		

      1mark
      (b)   By using trapezoidal rule, estimate the area bounded by the curve $y = 40 - 3x^2$, line x = 0 and line x = 2 using four strips of equal size.

10.   A quantity P is partly constant and partly varies inversely as square of t. P = 6 when t = 6 and

      p = 18 when t = 3. Find t when p = 11

11.   The position vectors for points A and B are 2i + 5j + 3k and 4i – 7j + 3k respectively. Express vectors AB in terms of unit vectors i, j and k. Hence find the length of AB leaving your answer in simplified surd form

12.   Two trains 246 metres and 304 metres long are traveling towards one another at 114km/ hr and 66km / hr respectively. How long do the trains take to pass one another?

13.   Given that $\log_4 x + \log 2^2 = \dfrac{5}{2}$. Find the product of the possible values of x.

14. Use binomial expansion to simplify

$$\left(\sqrt{2} + \sqrt{5}\right) - \left(\sqrt{2} - \sqrt{5}\right)$$

15. A trader mixes grade A coffee costing sh 600 per kg, with grade B coffee costing sh. 280 per kg in the ratio 3 : 5.Find the price at which he must sell 1 kg of the mixture to make a profit of 20 %

16. Calculate the equation of the normal to the curve $y = 3x^2 - 4x + 5$ at the point (1,4)

17. Every morning during class time, Pendo either reads a novel or solves Mathematics questions. The probability that he reads a novel is $\frac{4}{5}$ .If he reads a novel, there is a probability of $\frac{3}{4}$ that he will fall asleep. If he solves Maths questions, there is a probability of ½ that he will fall asleep.

Sometimes the teacher on duty enters Pendo's classroom. When Pendo is asked whether he had been asleep, there is a probability of $\frac{1}{5}$ that he will admit that he had been asleep and a probability of $\frac{3}{5}$ that he will claim to have been asleep when he had not been asleep.

Using a tree diagram

Find the probability that;

i)   He sleeps and admits it.

ii)   He sleeps and does not admit it

246

iii)   He does not sleep but claims to have been asleep.

iv)   He does not sleep and says that he has not been asleep.

v)   He sleeps and admits and changes his mind.

18.   The marks obtained by 10 students in a maths test were
25, 24, 22, 23, x, 26, 21, 23, 22, and 27

The sum of the squares of the marks, $\sum x^2 = 5154$

a)   Calculate the
   (i)   Value of x

   (ii)   Standard deviation

(b)   If each mark is increased by 3, write down the

(i)   New mean

(ii)   New standard deviation

19.   A, B, C and D are four points on the surface of the earth and N is the North Pole.

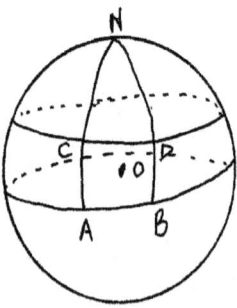

(a)   Calculate the distance, in nautical miles, between A (0°, 30°w) and B (0°, 60°E), measured along the equator.

b) An aircraft flies from A to B, then to N and finally back to A. Find the total distance traveled in nautical miles

c) Calculate the distance between C (55°N, 30°W) and D(55°N, 60°E)

d) Another aircraft travels from A due north to C, then due east to D and finally due south to B. Find in nautical miles the total distance traveled, and total flying time, given that the aircraft flies at a steady speed of 160 knots.

20.	In a single OAB, OA = a and OB = b. A Point C on OA is such that OC: OA = 2:5 D divides

AB in the ratio 2 : 1 OD intersect BC at X.

a)	Express OC , OD and BC in terms of a and b

b)	Given that OX – n OP and BX = m BC where n and m are scalars, express OX in two ways hence find the value of n and m

21.	The figure below is a pyramid of rectangular base ABCD of length 12 cm and width 9cm. The slanting edge has a length of 19.5cm

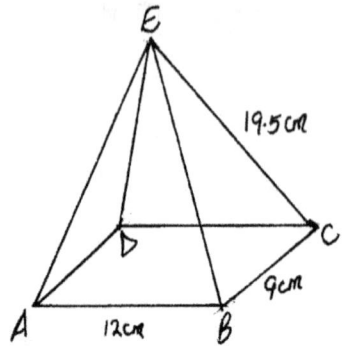

a)   Determine the height of the pyramid

b)   The angle AE makes with base ABCD

c)   The angle AED makes with BEC

d)   The volume of the pyramid

22.   a)   Using a ruler and a pair of compasses only, construct a triangle ABC such that
              AB = 8cm and angle ABC = CAB = 75°.

       b)   Locate the locus of P such that Angle APB = 52.5° and area of triangle APB = 20
cm² .

              Locate the two possible loci as P₁ and P₂. Measure the distance P₁ and P₂

       c)   On the opposite side of AB as construct a rectangular AB XY such that its areas is
48cm². Shade a point R inside the rectangle ABXY such that angle XRY > 90⁰ and angle
              ARB > 90⁰ locate and shade the region in which R lies.

23.   a)   Complete the table below for y = y cos x and y = sin (x +30)

x	-90	-60	-30	0	30	60	90	120	150	180	210	240	270
2cos x	0	1.0		2			0						
sin x + 30	- 0.87	-0.51		0.5			0.87						

       b)   Draw the graph y= 2 cos x and y = sin (x + 30) on the same axes.

              Use a scale of 1 cm to represent 30° on the x – axis and 2 cm for 1 unit on the y-
       axis.

c)   Using the graph solve $\frac{1}{2} \sin(x + 30) - \cos x = 0$

d)   State the amplitude and period of $y = 2\cos x$

24.   A manufacture has 720 and 840 kilogram of Arabica and Robusta coffee to make two brands of coffee A and B. To make brand A he mixes 60% of Arabica with 40 % of Robusta coffee. To make brand B he mixes 30% of Arabica with70% of Robusta coffee.

Brand A coffee to brand B must not be less than 3:5

(a)   Write down the inequalities representing the information above

(b)   Draw the inequalities on the grid provided

c)   Determine the maximum possible profit of the manufacturer if he sold each kilogram of A and B at a profit of Ksh 200 and Ksh 300 respectively. Note that the coffee was packed in tins of 1 kg only.

SOLUTIONS TO CHAPTER FOURTEEN

1.	$\dfrac{5.246 \times \log 0.2349}{0.06364 \, \frac{1}{2}}$

$= \dfrac{5.246 \times \ \overline{1}.3708}{0..6364 \, \frac{1}{2}}$

$= -\dfrac{(5.246 \times 0.6292}{0.063664 \, \frac{1}{2}}$

No	Log	
5.246	0.7198	
0.6292	$\overline{1}.7988$	
	0.5186	0. 5186 -
0.06364 ½	½ x $\overline{2}.8038$	$\underline{\overline{1}.4019}$
		1.1167

$$10^{-1} \times 1.309$$
$$= -0.1309$$

2.	$m = \sqrt{\dfrac{x-p}{p(1+px)}}$

$m^2 = \dfrac{x-p}{p(1-px)}$

$m^2(p - p^2x) = x - p$

$m^2p - m^2p^2x = x - p$

$m^2p + p = x + m^2p^2x$

$m^2p + p = (1 - m^2p^2) x$

$x = \dfrac{m^2p + p}{1 - m^2p^2}$

3.	CX.CD = AX. XB

$8 \times 5 = AX \times 4$

$AX = \dfrac{8 \times 5}{4} = 10cm$

4.	$6x^2y^2 - 4xy^3$

$9x^3y - 4x^3$

$= \dfrac{2xy^2(3x - 2y)}{xy(9x^2 - 4y^2)}$

$= \dfrac{2y(3x - 2y)}{(3x - 2y)(3x + 2y)}$

$= \dfrac{2y}{3x + 2y}$

5.	$D = \begin{pmatrix} 2 & 3 \\ -5 & 0 \end{pmatrix} \begin{pmatrix} 1-4 \\ 5 & 1 \end{pmatrix} - \begin{pmatrix} 3 & -1 \\ 4 & 2 \end{pmatrix}$ $\begin{pmatrix} 17 & -5 \\ -5 & 20 \end{pmatrix} - \begin{pmatrix} 3 & -1 \\ 4 & 2 \end{pmatrix} = \begin{pmatrix} 14 & -4 \\ -9 & 18 \end{pmatrix}$ det = (18 x 14) – (-9 x -4) = 216 inverse = $^1/_{216} \begin{pmatrix} 18 & 4 \\ 9 & 14 \end{pmatrix}$
6.	Original vol $= \dfrac{1}{3}\pi r^2 h$ New vol $= \dfrac{1}{3} x \dfrac{100.2\pi}{100} x \dfrac{100.4r}{100} x \dfrac{100.4}{100} x \dfrac{101.2}{100} h$ $= \dfrac{1.022\pi r^2 h}{3}$ Change $= \left(\dfrac{1.022}{3} - \dfrac{1}{3}\right)\pi r^2 h = \dfrac{0.022\pi r^2 h}{3}$ $\%\ change = \dfrac{0.022\pi r 2h \; x \; 100}{\frac{1}{3}\pi r^2 h} = 2.2\%$
7.	 Midpoint of L$_1$= $\left(\dfrac{5+3}{2}, \ \left(\dfrac{3-1}{2}\right)\right) = (4,1)$ Gradient of L$_1$ = $\dfrac{-1-3}{5-3} = \dfrac{-4}{2} = -2$

	Gradient of $L_2 = \dfrac{1-y}{4-x} = -2$ $1 - y = -8 + 2x$ $y = -2x + 9$
8.	$15, 17, 18, 19, 21, 21, 22$ $LQ = \dfrac{15 + 17}{2}$ U.Q $= \dfrac{21 + 22}{2}$ $\quad = 16 \qquad\qquad = 21.5$ Quartile deviation $= \frac{1}{2}(21.5 - 16)$ $\qquad\qquad\qquad = 2.75$

9.	

x	0	0.5	1	1.5	2
$y = 40 - 3x^2$	40	38.5	37	35.5	28

$= \frac{1}{2} \times 0.5 \left(40 + 28\right) + 2\left(38.5 + 37 + 35.5\right)$
$= \frac{1}{2} \times 0.5 \left(68 + 22\right)$
$= 72.5$ |
| | |

10.	$p = c + \dfrac{k}{t^2}$ $6 = c + \dfrac{k}{36}$ $18 = c + \dfrac{k}{9}$ $216 = 36c + k$ ----- (i) $\underline{162 = 9\,c + k}$ --------(ii) $54 = 27c$ $\dfrac{54}{27} = c = 2 \quad k = 144\Big\}$ when $p = 11$ $11 = 2 + \dfrac{144}{t^2} \qquad 9t^2 = 144$ $\qquad\qquad\qquad\qquad t = 4$

11.	$AB = \begin{pmatrix} 4 \\ -7 \\ 3 \end{pmatrix} - \begin{pmatrix} 2 \\ 5 \\ 3 \end{pmatrix} = \begin{pmatrix} 2 \\ -12 \\ 0 \end{pmatrix}$ $AB = 2i - 12j + 0k$ $|AB| = \sqrt{2^2 + (-12)^2 + 0^2}$

	$= \sqrt{4+144}$ $= 148$ $= 12.17$
12.	Total distance $= (246 + 304)$ metres $\qquad\qquad = 550$ metres Relative speed $= 180$km/hr $\qquad\qquad = \left(\dfrac{180\ x10}{36}\right) = 50m/s$ Time $= \dfrac{550}{50} = 11$ seconds
13.	$\log_4{}^x + \log_2{}^{2^2} = \dfrac{5}{2}$ $\log_4{}^x + 2\log_2{}^2 = \frac{5}{2}$ $\log_4{}^x + 1 = \frac{5}{2}$ $\log_4{}^x = \frac{3}{2}$ $x = 4^{\frac{3}{2}} = 8$
14.	$\left(\sqrt{2}+\sqrt{5}\right) = 1.\sqrt{2}^4\sqrt{5}^0 + 4.\sqrt{2}^3\sqrt{5}^1 + 6.\sqrt{2}^2\sqrt{5}^2 + 4.\sqrt{2}^1\sqrt{5}^3 + 1.\sqrt{2}^0\sqrt{5}^4$ $(\sqrt{2}+\sqrt{5}) - (\sqrt{2}-\sqrt{5}) = 4 + 8\sqrt{10} + 60 + 20\sqrt{10} + 25$ $\qquad\qquad\qquad = (4 - 8\sqrt{10} + 60 - 20\sqrt{10} + 25)$ $\qquad\qquad\qquad = 16\sqrt{10} + 40\sqrt{10}$ $\qquad\qquad\qquad = 56\sqrt{10}$
15.	$\cos t = \dfrac{3X600 + 5X280}{3+5} = 400$ $100\% = 400$ $120\% = \dfrac{120X400}{100} = 480$

16.	$\dfrac{dy}{dx} = 6x - 4$
	$at\ x = 1\ \dfrac{dy}{dx} = 2$
	$\dfrac{y-4}{x-1} \times 2 = -1$
	$y - 4 = -\dfrac{1}{2}(x-1)$
	$y = {}^-1/2\,x + 1/2 + 4$
	$y = {}^-1/2\,x + 7/2$

17.	
	$i)\ P(SA) = \left(\dfrac{4}{5} \times \dfrac{3}{4} \times \dfrac{1}{5}\right) + \left(\dfrac{1}{5} \times \dfrac{1}{4} \times \dfrac{1}{5}\right) = \dfrac{12}{100} + \dfrac{1}{100} = \dfrac{13}{100}$
	$ii)\ P(S \sim A) = \left(\dfrac{4}{5} \times \dfrac{3}{5} \times \dfrac{4}{5}\right) + \left(\dfrac{1}{5} \times \dfrac{1}{4} \times \dfrac{4}{5}\right) = \dfrac{12}{25} + \dfrac{1}{25} = \dfrac{13}{25}$
	$iii)\ P(\sim SA) = \left(\dfrac{4}{5} \times \dfrac{1}{4} \times \dfrac{3}{5}\right) + \left(\dfrac{1}{5} \times \dfrac{3}{4} \times \dfrac{3}{5}\right) = \dfrac{12}{100} + \dfrac{9}{100} = \dfrac{21}{100}$
	$iv)\ P(\sim S \sim A) = \left(\dfrac{4}{5} \times \dfrac{1}{4} \times \dfrac{3}{5}\right) + \left(\dfrac{1}{5} \times \dfrac{3}{4} \times \dfrac{2}{5}\right) = \dfrac{2}{25} + \dfrac{3}{50} = \dfrac{7}{50}$
	$v)\ \dfrac{13}{100} \times \dfrac{13}{25} = \dfrac{169}{2500}$

| 18. | (a)i) |

	$25^2 + 24^2 + 22^2 + 23^2 + x^2 + 26^2 + 21^2 + 23^2 + 22^2 + 27^2$ $= 5154$ $x^2 + 5073 = 5154$ $x^2 = 81$ $x = 9$ ii) Mean $= \dfrac{\sum x}{\sum f} = \dfrac{220}{10} = 22.2$ Variance $= \dfrac{\sum x^2}{\sum f} = - \left(\sum x\right)^2$ $= \dfrac{5154}{10} - 492.84$ $= 22.56$ $sd_x = \sqrt{var}\ x = \sqrt{22.56}$ $= 4.750$ b(i) New mean $\dfrac{222 + 30}{10} = \dfrac{252}{10} = 25.2$ (ii) New standard deviation 4.75
19.	a) $\angle d = 90°$ if $1° = 60$nm $90° = 60 \times 90$ $= 5400$nm

	b) Distance from B to N $\angle d = 90°$ $1° = 60$nm $90° = 60 \times 90$ $= 5400$nm c) btn c and d $\angle d = 90°$ $1° = 60 \cos 55$ $90° = 60 \times 90 \cos 55$ $= 3097.31$nm d) Distance ACDB Length AC $\angle d = 55°$ Dist $= 55 \times 60$ $= 3300$nm Total dist $= 3097.31 + 3300 + 5400 = 11797.3$ Time taken $\dfrac{11797}{160} = 73.73$ hours

20.

$OC \rightarrow 2/5a$ $OD = OB + BD$

$= \underset{\sim}{b} + 1/3\,BA$

$BA = BO + OA$

$= -\underset{\sim}{b} + c$

$OD = \underset{\sim}{b} - 1/3b + 1/3c$

$2/3b + 1/3c$

$BC = BO + OC$

$= \underset{\sim}{b} + 2/5c$

$OX = n\left(\dfrac{2b}{3} + \dfrac{1/3c}{3}\right)$

$= 4nb + \dfrac{nc}{3}$

$BX = m\left(\underset{\sim}{b} + 2/5c\right) = m\underset{\sim}{b} + 2/5mc$

$OX = O\vec{B} + B\vec{X}$

$\underset{\sim}{b} + m\underset{\sim}{b} + 2/5mc$

$\left(1 + m\right)\underset{\sim}{b} + 2/5mc = 2/3n\underset{\sim}{b} + n\underset{\sim}{c}$

$1 + m = 2/3n$ ......(i)

$2/5m = 1/3n$ ....(ii) $1 + m = 2/3 \times 6/5m$

$n = 6/5m$

$m - 4/5m = ^- 1$

$1/5m = {}^-5$

$n = {}^- 6$

21.

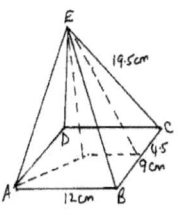

19.52 - 4.52 = 360 = 18.97

$\sqrt{18.97^2 - 6^2} = 17.996$

≅ 18cm

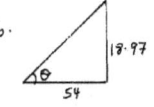

12^2 x 9^2 = 108 = 54

sin θ = $\dfrac{18.97}{54}$

θ = 20.57°

c)

E

18.97 / 18.97

Y — 12 — X

$12^2 = 18.97^2 + 18.97^2$ - 2 x 18.97 x 18.97 cos

144 = 719.722 – 719.722cos θ

cos = $\dfrac{719.722 - 144}{719.722}$ = 0.7999

θ = 36.88°

d) r = $^1/_3$ x 12 x 9 x 17.996

   = 647.856cm³

22.

B1 ∠75°

B1 ΔABC

B1Bisector of B/A

B1Centre

B1Locus P1

B1ocus p2

B1 Dist P1P2

B1Rectangle constructed

B1Bisectore of ∠ x and y

B1Region shaded

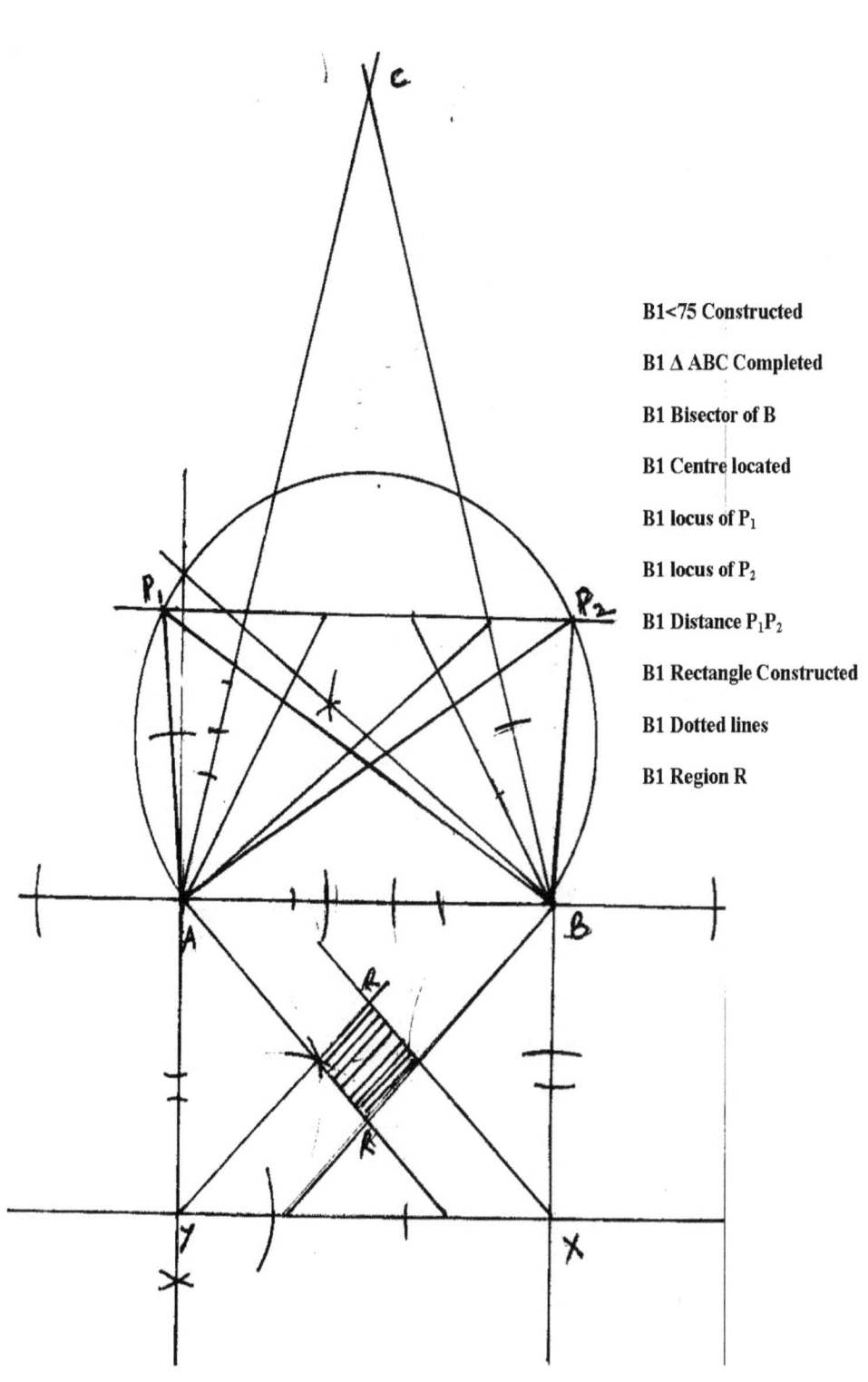

B1<75 Constructed

B1 Δ ABC Completed

B1 Bisector of B

B1 Centre located

B1 locus of P_1

B1 locus of P_2

B1 Distance P_1P_2

B1 Rectangle Constructed

B1 Dotted lines

B1 Region R

23	a)													
	x	-90	-60	-30	0	30	60	90	120	150	180	210	240	270
	2 cos x	0	1.0	1.73	2	1.73	1	0	-1	-1.73	-2	-1.73	-1	0
	sin x + 30	-0.87	-0.51	-0.87	0.5	0.87	1	0.87	0.5	0	-0.5	-0.87	-1	-0.87

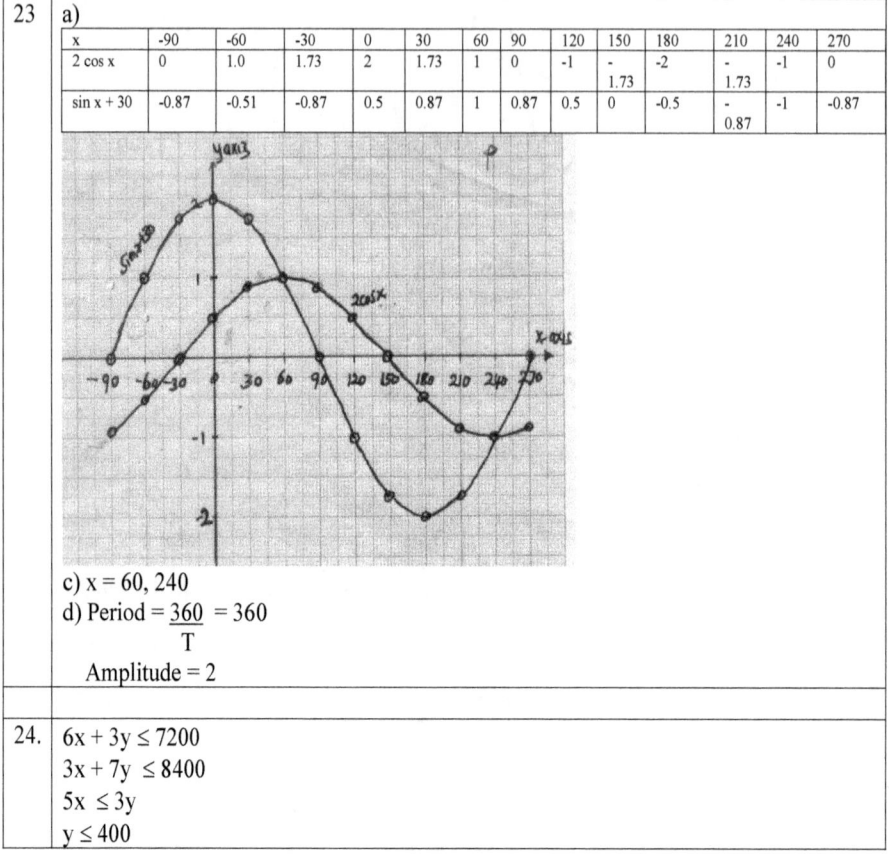

c) x = 60, 240

d) Period = $\dfrac{360}{T}$ = 360

    Amplitude = 2

24.	$6x + 3y \leq 7200$
	$3x + 7y \leq 8400$
	$5x \leq 3y$
	$y \leq 400$

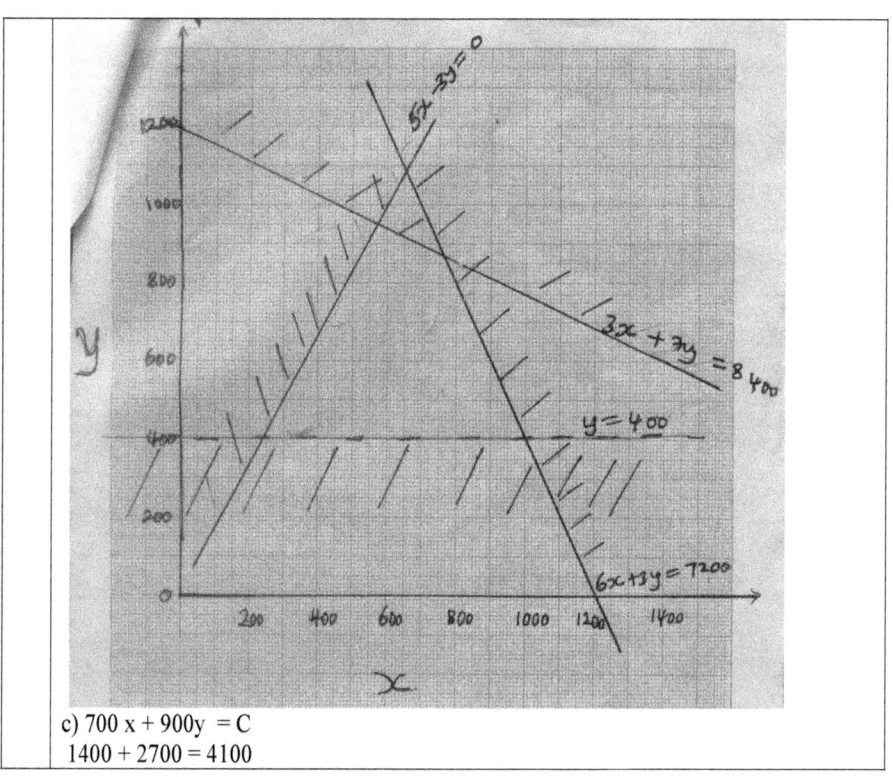

c) 700 x + 900y = C

1400 + 2700 = 4100

www.ingramcontent.com/pod-product-compliance
Lightning Source LLC
Chambersburg PA
CBHW051857170526
45168CB00001B/140